Connecticut River SHIPBUILDING

Wick Griswold & Ruth Major

Published by The History Press
Charleston, SC
www.historypress.com

First published 2020

Manufactured in the United States

ISBN 9781467144476

Library of Congress Control Number: 2020932104

For Annie and Maggie Mae the Bulldozer—Griswold

For Paris, Isabella, Gabriel and Tati—R.F.W.

Contents

Contents

Acknowledgements

Wick Griswold

There are so many wonderful people who provided me with insights, information and encouragement as this project came to fruition, and a heartfelt thanks goes out to all of you. First, I must strew roses in the path of the marvelous Melissa Josefiak, the executive director of the Essex Historical Society. Melissa provided the impetus for this book and has supported its development from initial research to completion. Her expertise, warmth, energy, professionalism and spirit of fun are inspiring. Brenda Milkofsky, the Kahuna of all things Connecticut River, was an important source of information. Amy Trout, curator of the Connecticut River Museum, was, as always, generous with her time and collections. Geoffrey Paul, the keeper of the Griswold Inn (and all things Essex), was so very generous sharing his art, wisdom and love of history. Jack Nelson, noted Viking, was an excellent research assistant at Mystic Seaport. My mentor, kindred spirit and dear friend Stephen Jones was a source of inside information and editorial assistance. Ned and Carol Libby, along with Strick Hyde, could not have been more forthcoming and helpful. Frederick Schavoir was a super source of knowledge, insights and encouragement. Doug Domainie at Dauntless Shipyard was most generous with his historic files. Tim Eastland of Eastland Yachts went above and beyond with all the excellent images and information with which he graced me. The mega-talented Caryn B. Davis was just splendid in terms of sharing images and ideas. Noted polymath Fred Frese's firsthand recollections were fascinating and fun to hear. The photography of Liz Burnell was spot on. Dani McGrath and Mike Kinsella from The History Press were terrific. It is a great pleasure to work with Ruth Major; her willingness to share her family history is just awesome. As always, it took a village, and what a wonderful village it is. Thank you, one and all!

Ruth Major

Allow me to add a few more praises to this village of historical book contributors. First and foremost, I wish to recognize the knowledge that Shirley and Don Malcarne have shared over the past twenty-five years since I first wrote to them. Don's notes, books and articles are invaluable, and Shirley's educational tours of the three Essex villages, Essex Historical Society and the Connecticut River Museum—and dinners at the Gris—are days my daughter, Paris, and I cherish. Appreciation and gratitude go out to Jim Powers and Fred Szufnarowski for their willingness to coordinate Jim's walking tour that included Meadow Woods, Williams Shipyard and the probable site of New City yard. I am still awestruck from Geoff Paul's tour of Captain Champlin's home and thank Geoff, Amy Trout, Melissa Josefiak, Rhonda Forristall, Paris Major, Jessica Reeves, Judy Micoleau, Marjorie Whitney and Captain T.A. Canham for sharing photos, knowledge and time. I am grateful to Paul Kilpin, master shipwright, educator, draftsman and my technical advisor on building and launching wooden ships. A special hug goes out my daughter, Paris, who shares research, historical road trips and life with me. Thank you to Brenda Milkofsky for sending facts on the ship *Orphan* and to Reverend Dr. Jonathan H. Folts, who found historical data regarding early rectors of St. John's Church in Essex and approved of the *Orphan* launch and blessing. Last but not least, I want to thank History Press editor Mike Kinsella, sales representative Dani McGrath, production editor Abigail Fleming and my co-writer, Griswold cousin and master wordsmith Wick Griswold for their trust and guidance. This book truly is a collaborative effort, and we are fortunate to have such wise, amiable and accommodating people contribute to the production of *Connecticut River Shipyards*. With sincere appreciation, thank you.

Introduction

A stroll down Essex, Connecticut's Main Street is a saunter back into the textured history of the United States of America. In just a quarter of a mile, one can amble along a corridor of time that stretches from the deep past into the ever-changing present and future. Every May, this storied street becomes a parade route that celebrates a seminal event from the town's, and our country's, early annals. Fifers and drummers dressed in the blue-jacketed, brass-buttoned, striped blouse regalia of early nineteenth-century sailors march in musical cadence to celebrate one of the most important naval non-battles in U.S. history. They step smartly past stately homes and one of the oldest taverns in the country as they head downhill to the waterfront. It was there that British marines landed in 1814 and burned the fleet of American privateers built to menace English shipping.

Today, at the spot where the British marauders landed, stands the Connecticut River Museum at the old Steamboat Dock. Tied up to the wharf in front of the museum is an excellent re-creation of Adriaen Block's ship *Onrust*. It was the first wooden sailing boat built in North America. When his fur-laden trading vessel, the *Tyger*, was set ablaze under mysterious circumstances off Manhattan Island in the fall of 1613, Block, his crew and their carpenter constructed the *Onrust* out of material salvaged from the burned-out hulk and wintered on Manhattan. Early in the spring of 1614, Block and his crew sailed east through a nasty stretch of water they named Hellgate and made their way up Long Island Sound on a voyage of discovery.

The *Onrust* was the first vessel manned by Europeans to sail up what was to become the Connecticut River. Block named it the Versch, or Fresh River, but the Algonquin name is the one that has come down to us and lasts though time. Imagine the awe the vision of that mysterious, white-sailed ship must have inspired in the Nehantic natives who inhabited the mouth of the river. It could move without paddles! These intrepid Dutch explorers sailed, rowed and kedged sixty miles upstream until they reached a set of rapids that blocked farther progress. Block turned about, came back downstream, interacted with a few groups of indigenous people and set the stage for European colonization of the river, its valley and its people. The Connecticut River would never be the same.

The Dutch were eventually supplanted by the English, who settled the length of the river from the Enfield Rapids to Long Island Sound. They started at a fort built at the mouth of the river and settled upstream in what was to become known as Potopaug Quarter. It encompassed the current communities of Essex, Old Saybrook, Chester, Deep River, Centerbrook, Ivoryton and Westbrook. These communities are the stage on which our narrative unfolds. They developed into shipbuilding and shipping enterprises that filled the world's waterways with Connecticut-made vessels, produce and goods. From the seventeenth to the twentieth century, some of the most beautiful sloops and schooners ever built came from these towns.

Today, Essex and its surrounding environs are marked by stately houses, upscale businesses and yacht repair and storage yards. It is a community that is most aware of its heritage. Stepping back in time is not a cliché here. The sense of the past made present is palpable in the carefully preserved architecture. The collection of maritime art that graces the walls of its most famous watering hole, the Griswold Inn, is world famous. The relaxed ambience of the village is a special quality of life for those who live and visit this historic hamlet. The story of its maritime heritage is one that takes us from colonization to revolution and on to the ages of sail and steam. Essex-built boats played roles in the Civil War and the First World War. The town's waterfront history encompasses the gilded age of yachting and is still a boating mecca today. Like the Connecticut River, Essex flows into the future with a strong sense of the past.

The Early Days

As later discussed in our look at dugout canoes, humans occupied the area that became Essex for thousands of years. After Adriaen Block's initial exploration of the Connecticut River revealed its seemingly unending supply of beaver, mink, otter and fisher cat pelts, entrepreneurs from Holland were quick to follow in his wake and establish trading relationships with the indigenous inhabitants. Some of these interactions were less than fair, and tensions between Europeans and the native people were inevitable. But it was conflict with the English that spelled the end of Dutch trading on the river and the ascendency of Puritans from Great Britain.

The first Dutch traders nailed their national coat of arms to a post on the west bank of the mouth of the river, a place they called Kviet's Hoek, Plover's Corner in English. The Hollanders then proceeded to establish a trading post/fort at what is now Hartford. They called it the House of Good Hope. It sat at the confluence of the Connecticut and Park Rivers. That fort/trading post became the hub of fur and goods exchanged between the Dutch and the River Indians, who inhabited the area. The coming of Europeans to the Connecticut River Valley was the most significant transformative event in the history of the region. Perhaps the greatest cultural collision in human experience occurred when people who possessed writing, metalworking, plows and firearms came in contact with non-literate societies that functioned with organic technologies based on wood and stone and a nature-based world view. Plow agriculture, and the metallurgy it inspired, proved the pivot point from which modern civilization ensued.

The Dutch, aboard the *Onrust*, not only brought advanced weapons technologies, but they also carried kegs of rum, an intoxicant that the river-dwelling people had no familiarity with and no physical or social means to ward off its addictive qualities. The men from Holland also carried the concept of private property. In the foraging and horticultural lifestyles of the indigenous river dwellers, possessions were shared. The idea that someone could own the earth was incomprehensible. Rum, guns and land ownership gave the interlopers tremendous economic and political advantage and power. They also brought viruses and bacteria against which the Algonquin-speaking natives had no immunities. Thousands of people sickened and died; communities and cultures were devastated and destroyed, never to rebound or recover.

The Dutch were primarily traders and merchants. Their overarching goal was to obtain as many beaver, and other animal, pelts as quickly and as cheaply as possible and ship them back across the Atlantic and sell them for handsome profits. Europe drove its beaver population to extinction because the animal's fur could be fashioned into waterproof, warm and stylish hats that were in great favor among men of means. Adriaen Block, and his countrymen who followed in his wake, were able to realize tremendous profit margins as they took unfair advantage of the people who trapped and processed hundreds of thousands of animal skins.

But military and political events in Europe began to spill over into what was coming to be known as the New World. The English and Dutch were at loggerheads, and that antagonism was to cross the Atlantic and manifest itself on the Connecticut River. Shortly after the Dutch built their trading post/fort at what is now Hartford, the English set their sights on settling the banks of the river. They were agriculturalists more than merchants and came to the pristine riverside with the intention of developing permanent settlements rather than just extirpating the existing beaver population and sailing away in search of more skins. Their plows were as effective as their cannon in driving the Dutch off the river and subjugating the native population.

In 1633, Lieutenant William Holmes led a small band of Puritans from Massachusetts up the Connecticut River in hopes of establishing a base upstream of the Dutch that would allow the English to facilitate trade with the local trappers and establish a farming community. In order to do this, they had to sail past the Dutch fortification at the House of Good Hope and its cannons. Jacob Van Curler, the commander of the fort, hallooed across the water and commanded the English ship to turn about. He ordered his cannoneers to load their weapons and light their matches.

Martial drumbeats filled the air. The stage was set for armed conflict that could exacerbate hostilities in Europe. It was a tense situation fraught with international consequences.

But the steely nerved Lieutenant Holmes ignored the Dutch threat and sailed blithely upstream. Van Curler blinked. The matches were extinguished, and the English erected the first prefabricated house in North America in what is now Windsor, Connecticut. The English had come to the river to stay. And stay they did. The inevitable war between England and Holland broke out in Europe. Tensions and bad feelings between the Dutch traders and the ever-growing population of English settlers continued to fester on the riverbanks. Faced with overwhelming odds, the Dutch realized their position was untenable and, in 1653, shut down their outpost and sailed down the river and out into the sound, never to return.

But the Dutch were not the only hostile people that the English bumped up against on the Connecticut River. At the point where the river meets the sound, where the Dutch nailed up their national coat of arms, the military engineer Lion Gardiner established a fort that commanded the mouth of the river with "greate gunnes." The fort was named Saybrook, in honor of Lord Saye and Sele and Lord Brook. They were English noblemen who facilitated British immigration to the Connecticut River Valley. While the River Indians, for the most part, were prone to peace and accommodation with the Europeans, some indigenous people, particularity the Pequots, had no love for the English. The unfortunate result of the growing tensions was the Pequot War, which resulted in the almost total decimation of the tribe and the ensured dominance of the English in Connecticut.

After the successful completion of the campaign against the Pequots, Lion Gardiner retired to an island off Long Island that still bears his surname. Command of the fort passed into the hands of Colonel George Fenwick, who established an agricultural, trading and exploratory presence on what is now known as Saybrook Point. Fenwick was married to the magical Lady Alice Fenwick, who is still celebrated as a pioneering female presence in the New World. She befriended the native inhabitants; grew crops, flowers and medicinal herbs; sang madrigals; rode horses; sailed small boats; and generally became an archetype for women transforming wilderness into civilization.

Her husband played a key role in transforming wilderness into habitation for colonists. As more English settlers moved into the fort and its immediate surrounds, George Fenwick anticipated a time when settlement would have to branch off into hitherto unexplored nearby areas. A fire at the fort in the winter of 1647 made the need for more living space a reality. Although the fort was quickly rebuilt, Fenwick dispatched key members of his administration to explore and survey possible dwelling sites. William Pratt and William Hide were those selected to look into what soon became known as Potopaug Quarter. The name, which has various definitions in Algonquin, was passed to the English colonists by their guide, Joshua.

Joshua was the son of Uncas, a Mohegan chief who was closely allied with the English. Joshua's indigenous name was Attawanahood, and he was recognized as a chief of the Western Nehantics. The close bonds established with these leading members of the native community spared the English who settled in the area from Indian attacks that plagued so many of their fellow settlers upriver in the years to come.

Joshua's name lives on today. Just above Essex on the east bank of the river, there is Joshua's Rock. It is a piece of the earth that once was set in Avalonia off the coast of Africa. Shifts in tectonic plates placed it here, in Connecticut, for the present. There also is a Joshuatown Road on the east side of the river. Joshua thought so highly of William Pratt that he recommended him to provide guidance to his children upon his death. Joshua also willed parcels of land to other English colonists. Joshua led Hide and Pratt to a lush paradise of meadows and coves. These were ideal for agriculture and maritime activities. The area also included what is now known as the Falls River, which provided water power to turn saw- and gristmills and mineral deposits of iron to be fashioned into hardware and tools.

These early settlers were joined by the Lay brothers, John, Robert and Edward. All three worked closely with George Fenwick. They managed his farm on what is now called Nott's Island and also oversaw the day-to-day operations of the farm at Cornfield Point, which played an important role in the conflicts with the Pequots. Pratt, Hide and the Lays built houses in or near what is the center of modern-day Essex. They established corn and wheat fields, orchards and pastures. Agricultural production necessitated the development of ferries across coves and cart paths that evolved into the streets that continue to be traveled in the town today.

These early settlers of Potopaug picked a particularly propitious part of the river valley to establish their young community. In addition to the area having excellent arable land to facilitate the production of food, the watercourse now known as the Falls River had a series of waterfalls that the colonists quickly utilized. By 1690, a gristmill was in operation at the Hough Dam. The financial arrangement that the residents worked under stipulated that the miller could turn a profit, but he also had to grind corn and wheat for his neighbors. This reliable food source assured the new residents that their relationship to the territory would be sustainable.

The damming of the Falls River set the stage for the interrelated activities that were to merge into the shipbuilding industry for which the area became famous. In 1697, peat bog iron ore began to be worked near the river. The early colonists in Connecticut were dependent on England for metal products in the colony's early days. The settlers brought the tools and hardware they required across the Atlantic with them. The need for hardware to build both houses and ships, as well as the machines of the nascent Industrial Revolution, impelled the newcomers to seek their own sources of iron.

In 1701, ironmaster Charles Williams (possibly a descendant of Roger Williams) was recruited from the Rhode Island colony to develop the ironworks that ultimately led to vigorous shipbuilding in Essex. A triphammer built on the Falls River made the manufacture of ship hardware possible. In 1648, shipbuilding began on the Connecticut River upriver from Potopaug at Wethersfield. A group of citizens banded together to create the sloop *Tryall*. It was outfitted for a voyage to Barbados under the command of Captain Greenfield Larrabee. This venture turned a tidy profit, trading agricultural products for rum and sugar. The *Tryall*'s first voyage established trade between Connecticut River ports and the Caribbean that proved to be so successful that it lasted for over two hundred years.

Potopaug was quick to follow in Wethersfield's wake. In the 1650s, Robert Lay built a wharf at the present site of the Connecticut River Museum. Lay was instrumental in establishing the town in the West Indies trade. In 1666, he headed a group of local men who purchased shares in a New London–built ketch named *Diligence*. He and his partners worked out an insurance-like agreement for the voyage, focusing on whether or not a horse would arrive alive and, if it did, how much rum and sugar to take in trade for it. The *Diligence* also carried corn, grain, peas, hay and candles to the islands. The voyage was a success and became the precursor of many successful ventures from the town of Essex. It solidified the bond between the town and the sea that was to see over five hundred Essex-built ships sail down the Connecticut River to ply the waters of the world's oceans and bring fame, elegance and prosperity to the community. This is the story of those ships and the families who built and sailed them

After the Golden Age of Sail was supplanted by steam, Essex became a stop for the opulent steamboats that ran up and down the river. At one point, a passenger could embark in Hartford at 5:00 p.m., have dinner, listen to a small orchestra, enjoy a cocktail, sleep in a stateroom and wake up in New

York City. The steamers were eventually made redundant by the railroads and automobiles. But Potopaug remained true to its heritage and reinvented itself as place where some of the finest yachts to ply the world's waters were built and moored. The story of these vessels is a splendid coda for the town's centuries of shipbuilding.

Dugout Canoes

For thousands of years, people circled campfires on the banks of the Connecticut River and passed sacred stories from one generation to the next. The exploits and adventures of divine spirits and beings were vividly recounted in timeless tales and sagas. One interesting spiritual eminence was the Noah-like rabbit-trickster named Manabozho. He was the son of Mother Earth and the Wind God. He performed several key feats that allowed the indigenous people of the deep past to live and flourish in harmony with the beautiful river and land. Like Prometheus, he purloined fire from the divinities and passed it on to humans. He also gave people the gift of culture, which allowed them to experience a learned, shared way of life that could be transmitted from one generation to the next.

From a riverine perspective, this divinity's most important deed was saving the world when the Great Deluge threatened to sweep away all life in a raging torrent that looked like it would drown the world. Manabozho responded to this catastrophe by loading all the plants, animals and people onto a massive tree. On this woody ark, the inhabitants of Earth rode out the watery spate and were able to resume their accustomed lives when the flood receded. Manabozho then resumed his trickster ways and went about doing a bit of good here and a bit of mischief there. As a result of his divine intervention, the ecosystem was saved, and life could proceed apace.

The story of the deluge-defying tree conjures up a couple of interesting interpretations. The story of a Great Flood can be found in many of the world's mythologies. One of the most familiar is that of Noah and his ark. Many interdisciplinary scholars today are positing the idea that Noah's flood and many others were caused by rapid and extreme climate change. As we shall see, climatic variations played a role in the sociocultural evolution of the indigenous people of the Connecticut River Valley. Their lifestyles reflect adaptation to the ever-changing environment in which they lived.

The other analogy based on Manabozho's salvation story concerns the massive tree on which he placed all life to ride out the destructive deluge. It is but a small leap of the imagination to see that tree as a symbol of the dugout canoes that were a prime tool in the River People's adaptation to the environment of the Connecticut River, Long Island Sound and beyond. The dugout was an important means of food production, seasonal migration, communication and trade. It played vital roles in human activity on the river for several thousand years. People still paddle the currents and coves of the Connecticut in canoe-style boats. The river has seen many types of vessels arise on its banks to venture forth to the oceans of the world, but the canoe has outlasted them all.

Shipbuilding on the banks of the Connecticut River conjures mental images of sturdy sloops and sleek schooners eager to ply the world's oceans in search of adventure, profit and the prizes of privateering. Hundreds of sailing ships were built in Essex and its surrounding towns from the seventeenth to the nineteenth century. But before the arrival of the Euro-Americans with their tall masted ships with billowing sails, uncounted thousands of dugout canoes were crafted by the Nehantic and Wangunk people. Indigenous people lived in the Essex area for almost fifteen thousand years. In the four hundred years since Adriaen Block first navigated past the point of land called Potopaug, many different types of vessels have sailed the river. Opulent steamboats, oil-soaked tankers, Civil War gunboats, splendid yachts, icebreakers and more worked their way from Long Island Sound to Hartford and back. But the humble canoe remains long after the steamers, towboats and barges have vanished.

Called *misoon* in the Algonquian language, the dugout canoe was the vessel ideally adapted to the waterborne transportation needs of the foraging people who quickly came to inhabit the lower Connecticut River Valley after the last ice age receded. The early inhabitants left no written records, but archaeological and anthropological evidence documents their evolving lifestyles for thousands and thousands of years. Their sustainable nutritional strategies of fishing, hunting, trapping, scavenging and gardening each utilized canoes to procure and/or transport foodstuff, skins, tools and people. The dugout canoe, usually crafted from tulip poplar, was an organic vessel that was an integral link in their interconnected, nature-centric systems that connected people, water and land. Sailor/author Matthew Goldman describes a dugout canoe as "a tree with a sense of adventure."

The importance of the dugout canoe in Connecticut goes beyond its food-gathering functions. The canoe was the means by which trading networks spanning North and Central America operated. Canoes played a key role in the patterns of immigration that spread humanity from the Pacific to the Atlantic Ocean. They linked the people living up and down the Connecticut River and Long Island Sound with the rest of the continent. While groups, what we think of as "tribes," lived on different parts of the river, their primary identity was aligned with the land but the river. They collectively became known as the River Indians, and it was the dugout canoe that made that identity possible.

Paleo people were living in inland regions of Connecticut some twelve thousand years ago. Large mammals such as mastodons, saber-toothed tigers, giant beavers and sloths and horses roamed the land, which was essentially tundra. By 8000 BCE, these large animals had largely died out, mainly due to a continuing period of climate change; it is not known what role humans had, if any, in their extinction. These early residents lived in small nomadic kinship groups and moved frequently. They preferred to camp on elevated land that overlooked water sources. Sea levels were lower, so the Essex area was farther inland than it is today. But the life-giving waters of the river were, by necessity, always close by.

The climate had warmed to the extent that new varieties of plants and animals appeared in the region. The people who lived there now hunted deer, caribou, elk, bear and birds. These were not only food sources but also provided tools, clothes and sometimes shelter. Fish and shellfish became more accessible. Nuts and other land-grown, plant-based foods also increased, as well as edible water-borne plants. As more food became available, populations grew slightly, but they still remained essentially fluid and nomadic. Moving with the seasons, they headed inland in the fall and winter, shoreward in the spring and summer.

The Paleo people adapted and transformed materials found in nature into tools. Scrapers, hammers and knives were fashioned from stones and bones. Wood was utilized in a variety of ways. The opposable thumbs of the early residents came in handy in the development of strategies to increase food production. One of the key elements that separates humans from other animals is, of course, the creation and control of Manabozo's gift of fire. Fire made cooking possible, which decreased the chances of foodborne illness. It provided warmth and light and some protection from predators.

Fire, along with cutting and scraping and tools enabled ingenious early inventors to burn out sections of fallen trees and then scrape them into concavities. One conjecture is that the first dugouts were the result of lightning-struck trees, partially burned, that were observed floating with something in them. The dugout was a huge improvement over merely straddling a log and paddling because it allowed humans to transport not only themselves but also anything else that might fit in the boat. And they were a lot less tippy. Some of the larger manifestations of these vessels were oceangoing and could hold up to forty people. They could navigate in open water up and down the Atlantic coast and go inland on major river systems.

Ongoing climate change continued to challenge the early inhabitants of Connecticut. As a result, the Paleo people were eventually replaced with the Archaic people. An archaeological site just across the river from Essex dates back 9,500 years. One of the intriguing aspects of this site is that some of the implements discovered there were not of local origin. The presumption is that they were either traded or carried by the inhabitants from afar. Either scenario suggests that canoes were involved in their transport. Tools from the period were also found along Long Island Sound from Old Saybrook to Madison.

The Archaic people who replaced the Paleo inhabitants began to live in larger communities with more formal social organization. It is during this time that fishing became a vitally important means of harvesting protein. Saltwater fishing, especially, required people to be able to navigate over sometimes turbulent waters. The fact that sharks and other large fish were among the prey of early fishers indicates the existence of canoes that could carry people and cargo offshore. During this period, people began building fish weirs on the Connecticut River that would also require cargo-carrying canoes and deft seamanship skills. Archaeological evidence suggests that during this period, trade with groups up and down the river and sound was common.

Several shell heaps in and around the mouth of the river show that bivalves were an important food source of these people. Dugout canoes would have been vital in transporting gatherers to the oyster, clam and mussel grounds. The canoe would also carry the piles of the delicious creatures to where they could be shelled and shucked. Lobsters and crabs could also be captured with the help of dugout canoes. Fish weirs of wood and stone enabled people to corral the massive migrations of salmon, shad and herring that teemed up the river every spring. Again, the dugout canoe made the building of these weirs possible and were important in transporting the captive fish back to shore for processing.

Around three thousand years ago, the Archaics began to be replaced by the Woodland people. This group survived up until the time that Europeans sailed into the river and changed ways of life forever. This era introduced the use of clay pottery and pipe smoking. Pottery enhanced the storage of food and water. Pipe smoking became a numinous ritual activity, primarily for men. Tobacco was seen as sacred, and a mixture of it along with some other plants and herbs called *kinnikinnick* was used in a variety of spiritual and secular activities. The practice and the shape of the pipes are believed to have come from the Midwest. Bow and arrows also developed during this period. This innovation proved more effective in the hunting of deer and elk. Dugouts were used to carry meat and skins up and down the river.

The importance of the dugout is elegantly stated by Verena Harfst in "A Brief History of Native Americans in Essex, Connecticut":

> *By the Early Woodland Period, the Connecticut River and Long Island Sound had become major trading routes, with dugout canoes going up and down the eastern seaboard, and also traveling up river further inland. The river did not constitute a boundary, but rather served as an artery of communal life and trade....Algonquin communities in New England and their languages developed separately, but the various communities thought of themselves as sharing a common culture.*

One thousand years ago, the canoe again played a pivotal role connecting Connecticut River People with the larger world. Via trade routes that crisscrossed the continent by water, new foods and technologies arrived in the river valley. Horticultural techniques and the seeds and means to cultivate corn, beans and squash, known as the Three Sisters, were introduced to the region. Using wooden digging sticks and clamshell hoes, the early farmers increased the availability of food. Canoes were critical in carrying seeds and cultivated crops from one place to the next. They also supplied the fish that were used as organic fertilizer.

For the most part, the River People lived a cooperative and harmonious lifestyle. Groups occasionally had their differences, but armed conflict was rare. One can look back on this time and grow nostalgic for a world that may never be seen again. It was from dugout canoes that the Nehantics of the Essex area first saw the prow of the *Onrust* cutting up the river on an

April morning in 1614. Their world would never be the same. Guns, plows, money and disease would change their ways of life forever. But, on a crisp April morning, you can still carry your canoe down to the landing at the end of Main Street and paddle off into the mist and calm water and imagine a beautiful world long gone.

Early Shipbuilding in Essex

Some families leave an indelible impression on their community. From the early eighteenth century through the nineteenth, the Hayden family of shipwrights, carpenters, figurehead carvers and ship merchants helped build, maintain and support the village in which they lived. The family left a legacy passed down through generations of expertly built ships and shipyard operation. Their knowledge, skills and traditions were shared with artisans who worked alongside the Haydens in their extensive shipbuilding and repair yard. A shipping and shipbuilding industry developed in the village, and several Hayden shipwrights and nearly all members of the community had their hand in it. The Haydens' legacy in Essex continues to inspire new generations of sailors, boat builders, restorers and merchants.

Uriah Hayden was born on January 10, 1732, and grew up in Potapaug. His father, Captain Nehemiah Hayden, taught Uriah the art and craft of boatbuilding. When Uriah was ten years old, Nehemiah built a two-masted snow, or "snaw," in the yard beside his home, as was quite common at the time. This type of vessel has a snow or trysail mast behind the main mast. Captain Hayden used his snow for years in the lucrative West Indies trade, established in the second half of the sixteenth century, and Uriah followed his father's footsteps. Beginning in 1750, Uriah built his own vessels in Potapaug, as did his cousin Ebenezer Hayden.

The fastest way to travel at the time was by boat, as roads were scarce and treacherous in foul weather. Coastal shipmasters who traded with merchants in the West Indies needed local wharves, warehouses and docks to unload and store merchandise and refill their vessels with trade goods. Abner Parker was granted a twenty-foot-wide section of the riverbank at Potapaug Point to build a wharf and warehouse in 1753 for "his convenience" and benefit of the general public. Parker built his wharf that year, and the following year, he constructed his warehouse. Parker's wharf and warehouse were essential components required by the community to increase their trade and commerce and provide storage and sales at one location.

Large stores of food and goods for the village and outlying areas were kept in the warehouse. Products included tobacco, lumber, salt pork, rum, sugar, molasses and pearl ash, used for making gunpowder, soap and as a leavening product in baked goods. These and other popular items were readily available for purchase and sale from Parker's warehouse to merchant mariners from other communities and the local villagers. In time, Parker's Wharf was owned by the Haydens and, later, became the steamship landing.

Some of the first residents of the village were Parkers and Tookers, who changed their name to Tucker. They were among the first shipbuilders in Potapaug. John Tooker was building ships on the Middle Cove as early as 1720, and Noah Tucker built ships up to 1750. Uriah Hayden purchased former Tooker and Parker properties on the south side of lower Main Street. About 1755, he built a home off one end of an existing Parker house. The construction crew dug into the riverbank for placement of the new larger front section of the Hayden house, and Uriah's walkout-basement conveniently faced the river.

Ann Starkey, sister of master shipbuilder Timothy Starkey, married Uriah Hayden and worked with her husband to design the new building. The enterprising couple decided to turn their large basement into a tavern. They had a sign made in England that hung conspicuously outside the tavern door and featured a full-rigged ship and "U & A, 1766" painted on the lower front section. The initials stood for the proprietors, Uriah and Ann. The couple's brass George the Third doorknocker could be seen, still shining, on the front door in 1874. The Hayden homestead remained in the family for many years. It changed hands several times, but in 1918, the place took on its current name, the Dauntless Club, a private yacht, sports and entertainment club and wedding venue.

In 1760, Elijah Scoville built a house on a country road in the village and sold it to Uriah's cousin Ebenezer Hayden. Ebenezer married his neighbor Prudence Pratt, and the couple moved into their new house on a dirt road that eventually became the fashionable West Avenue. Like his cousin Uriah,

Ebenezer Hayden dreamed becoming a successful maritime merchant. And he possessed the intelligence and fortitude required to do so.

The year 1770 was a significant one in Uriah Hayden's life. He became the local militia captain and established the Hayden Shipyard on land to the right of his house on Middle Cove. Eight years earlier, he bought eight acres of land near today's Mack Lane from his cousin Ebenezer's father. Uriah set up his shipyard on the edge of the cove for ease in building, launching and repairing his vessels. He built a house for his brother Elias and, in time, another house for Elias's son, Calvin. They and several other family members were affiliated with the Hayden Shipyard that ultimately spanned several generations.

Exposed to the many types of work in the shipyard, young Hayden and Starkey boys honed their skills as carvers, shipbuilders, joiners, carpenters and merchants by watching, apprenticing and learning from the master carvers, shipwrights and carpenters. After a long, hard day's work in the shipyard on a hot summer's eve, the Old Ship's Tavern was no doubt a popular watering hole for both yard boys and men. Ann Hayden kept tabs by placing pegs into holes carved into the tavern woodwork.

The marriage of Uriah and Ann was not the only Hayden-Starkey union in the village. Young merchants and brothers Timothy Jr. and Alex Starkey married Captain Uriah and Ann Hayden's daughters Mary Ann and Esther. It was not unusual at that time for first and second cousins and family friends to marry. Unions with close family ties served to strengthen and solidify extended families. Business partnerships between brothers and brothers-in-law were common and usually beneficial to families of the partners. "Keeping it all in the family," Marjorie Post would say, "is both prudent and practical."

In 1773, the proprietors of Potapaug gave permission for Captain Hayden to fill in between his and Captain Parker's 1753 wharf. This fill work expanded Hayden's shipping operation, and it increased the landing area and length of the dock. This allowed for more trade vessels to dock and unload at the same time. The expansion of the Hayden Wharf and dock caught the eye of the Connecticut's government officials, who needed a large wharf and dock.

In early 1775, colonists were tired of taxation by the British without representation, and talk of a revolution spread across New England in shouts and whispers. Proud Patriots raised liberty poles in town centers and caused British officials to rush out and "read the Riot Act" or grab a few rabble-rousers by the scruff of the neck and toss them in the gaol house.

Connecticut colonial governor Jonathan Trumbull was attuned to the situation and kept close watch on the shipyards. He knew the political climate grew more adversarial every day. As a sympathizer to the Patriot cause, like Governor Nicholas Cooke of neighboring Rhode Island, Governor Trumbull knew he had to proceed with caution and act judiciously. Together with the Council of Safety, a decision was made to acquire a warship. Facing the might of the British navy, the governor received authorization from the General Assembly to purchase and outfit a warship for the defense of the colony.

In 1775, Captain Uriah Hayden completed his new sail loft and warehouse. At forty-four years old, he had made a name for himself as a prominent shipbuilder and yard owner, and his tenacity and hard work were about to pay off. In January 1776, Uriah received an unexpected commission that would not only bring tremendous fame to himself and his shipyard but also denote Essex as a prime center for shipbuilding and mercantile activity in the lower Connecticut River Valley.

Jonathan Trumbull served as governor during the tumultuous transition period when Connecticut colony became the State of Connecticut. Known for his ability to mediate disagreements, Trumbull's lengthy tenure as governor of the colony ran from 1769 to 1776. From then until 1784, he was the first governor of the State of Connecticut, succeeded by Matthew Griswold, a multifaceted Patriot from Lyme. Trumbull selected Uriah Hayden to build Connecticut's first warship.

Hayden's assurances to Governor Trumbull included the usual ship designs, drawn by Captain Hayden, and a half-hull model for his master ship carpenters and laborers to use. They also included a commitment to secure materials for building the vessel and a written agreement that the ship would be completed by June 1776. The warship project required subcontracts from the colony to reputable adjunct businesses. One of the subcontracts went to Benjamin Williams for creating ship's hardware. This ensured profits for him and steady work for his ship smiths. Benjamin Williams set up his forge at a dam on the Falls River about 1761, and he was quite accomplished by the time his contract from the state arrived. Williams was responsible for turning seven tons of raw iron into the fittings required for the colony's new warship.

Additional contracts for materials to build the colony's warship were given to businesses to provide equally essential materials. Contracts were awarded for rigging, muskets, cannon and gun carriages. Uriah Hayden then established a chandlery near his home where supplies could be stocked and readily available. From January to June 1776, the Hayden Shipyard was fully engaged building the colony's biggest and most formidable warship.

Hayden's prestigious commission mustered much excitement in the sleepy little village. It certainly identified Potapaug as the place to build and buy ships, and it promoted a period of prosperity, growth and activity in the community. This was not only true for those who worked at the Hayden Shipyard but also for all the ancillary craftsmen around town who provided necessary materials for ships, such as sails, ropes of several widths, spars and trunnels, or tree nails. These were large nails crafted out of wood and used to secure timbers. Carvers were needed to carve figureheads, billet heads and ship models. Block and spar makers, riggers, awlers and many other skilled artisans whose livelihoods depended on shipbuilding were kept busy creating parts for the *Oliver Cromwell*, as the ship would be known.

During the American Revolution, Abner Parker's warehouse was filled with mounds of salt. Without a means of refrigeration aboard ships, this item was an essential commodity that preserved food for sailors at sea, and soldiers on land also relied on it. Salt was used to cure and preserve leather goods required of farmers, local men, cavalry and mariners.

Uriah Hayden's warship project undoubtedly promoted much interest and pride in the hearts of folks who resided in the colony, but especially for those who lived and worked in Potapaug. In the spring of 1776, Hayden's project and the increased demand for goods and services it generated were the talk of the town.

Captain Seth Harding was assigned to oversee the vessel's construction by shipwright Uriah Hayden. The *Oliver Cromwell* was a fine ship—three hundred tons displacement, fully armed and loaded with men and provisions. Its keel was eighty feet long, and it had a twenty-seven-foot beam and twelve-foot draft. At the time, *Oliver Cromwell* was the largest full-rigged ship in the colony of Connecticut. Visiting officers and dignitaries rushed down to Hayden's dock on June 13, 1776, to witness and celebrate the launch of the great ship. It slid down the ways and splashed into the cove to the sounds of whistles, clapping and shouts of "Huzzah!"

Captain William Coit commanded the new ship, but within a few days of the *Cromwell*'s launch, a violent lightning storm damaged two of its masts. Significant repairs had to be made. Two months later, Captain Coit finally sailed off for New London, where the ship was fully outfitted with stores and arms. A series of delays and changes in personnel ensued. By the time the ship was inspected by the governor in April 1777, *Oliver Cromwell* had twenty-four or more mounted guns, the ability to carry 180 men and a new commander. Captain Coit was removed, and Seth Harding replaced him.

Coit's accounts with Uriah Hayden, Benjamin Williams and others were settled, but not without some fractious acrimony.

Captain Seth Harding was well suited to command the *Cromwell*. In the summer of 1778, while en route to France from Charlestown, the *Cromwell* was hit by a devastating hurricane. Captain Parker had no other option than to jury rig the ship and head back to New London for repairs.

On June 6, 1779, while in company with the privateer *Hancock* some leagues off Sandy Hook, New Jersey, *Oliver Cromwell* was surrounded by three British warships and a brig. *Cromwell*'s crew eventually took out the main topmast of one of the British ships, but despite its valiant engagement with the other British vessels, it was outnumbered and ultimately defeated. Captain Harding, his officers and crew were imprisoned but eventually given liberty to leave.

The British renamed the Connecticut colony's largest warship *Restoration*, but they could not take back the prizes it had won between 1777 and 1779. The *Oliver Cromwell* took nine valuable prizes in just two years. These included a brigantine, *Honour*; a brig, *Medway*; ships *Restoration*, *Admiral Keppel* and *Weymouth*; the sloop *York*; and schooners *Dove*, *St. George* and *Hazard*, one of which was filled with desirable mahogany and log wood. Surely these prizes compensated somewhat for the loss of *Oliver Cromwell*.

There was no substantial American navy at the beginning of the American Revolution, as it had yet to be established. However, there were many enthusiastic young men from Saybrook villages manning privateers and eager to confront British ships and protect their communities. These local vessels significantly impeded the British West India trade. Two thousand American vessels, mostly sloops and schooners, were at sea during the Revolution. Some were able to disrupt British supply lines and raise havoc with their attempts to get provisions and supplies to their forces in New York and Long Island. Spoils taken from captured British prizes were sorely needed by Connecticut colonists and ameliorated the shortages of supplies available to both the citizens and the military. A warship like *Oliver Cromwell* provided an effective means of attacking British ships at sea, but armed privateers also made significant contributions.

Nathan Post, a descendant of Saybrook settler Steven Post, fought at sea and commanded three armed Connecticut vessels. These included the sloop *Revenge*; a brigantine, *Delight*, armed with eight guns and carrying twenty men; and, in 1782, the brig *Martial*, mounted with sixteen guns and carrying eighty men. Captain Post was credited with helping cut off the British supply route from New York to the West Indies. While this certainly was a significant accomplishment, Captain Post was but one of many patriotic Connecticut shipmasters and mariners who served the colony in armed and unarmed vessels. Their boats were positioned all along the Eastern Seaboard, at the mouths of rivers and out at sea during the lengthy battle for independence.

The Turtle

Like many good sea stories, this one begins in a bar. One evening in 1974, as America was approaching its two-hundredth birthday, a polymath named Fred Frese and a buddy of his were enjoying a cocktail at a venerable New Haven watering hole. The conversation somehow meandered around to the realization that they were close to the spot where, two hundred years in the past, another polymath named David Bushnell had a conversation with a multifaceted entrepreneur named Isaac Doolitle. Both of these eighteenth-century men were ardent supporters of the Patriot cause and talked long into the night about how they could best advance the cause of freedom that was sweeping Connecticut.

From that long-ago tête-à-tête grew the idea that became the most innovative boat ever constructed on the banks of the Connecticut River. It was the first submarine built to engage in warfare. The inventive genius of David Bushnell teamed with the mechanical ingenuity of Isaac Doolittle and Phineas Pratt (a scion of one of the founding families of Essex) combined to create, a singular, unique advancement, in terms of form and function, in the evolution of naval architecture. Their brainchild came to be known as the *Turtle*, although they didn't call it that. Its descendants became a whole new category of seagoing weaponry—a category that shapes the world's geopolitics today.

The story of the *Turtle* reaches from the Revolutionary War into the second decade of the twenty-first century. It is one of the most intriguing narratives of human ingenuity, courage and derring-do that can be found in our nation's history. Fred Frese's tavern-inspired conversation resulted in three reincarnations of Bushnell's marvelous engine of war. These re-creations can still be seen in hands-on exhibits at the Connecticut River Museum, and the USS Nautilus Submarine Museum in Groton, Connecticut. Should you find yourself in the vicinity of either (or both) of these excellent institutions, by all means, please stop in and experience the *Turtle* for yourself.

The ribbon that ties the *Turtles* together through time is the Connecticut River. David Bushnell built his vessel in the Westbrook section of Saybrook and tested it at Ayer's Point in Essex. Fred Frese and his team built their replica in Saybrook and tested it in Essex. It was the brackish water of the river, where the downstream flow of the current met the incoming tides of Long Island Sound, that gave both projects valuable insights as to how their vessels floated and submerged in varying levels of salinity. This information proved vital to Bushnell in his attempts to sink British warships in the Hudson River. It was also important to Frese as a piece of the important scientific information to include in his exhaustive study of all the relevant factors that were compounded to give a complete understanding of as many aspects of the *Turtle* as possible.

The Connecticut River provided Bushnell with a safe environment in which to conduct his field trials. During the Revolutionary War, Essex was very strongly pro-liberty. Essex gave the inventor access to men like Phineas Pratt and Uriah Hayden. Hayden, of course was the builder of the warship *Oliver Cromwell*. Fred Frese believes that Hayden provided Bushnell with sections of "compass wood that would conform to the curvatures for his elliptical submarine hull."

Bushnell was known to occasionally raise a glass at Uriah Hayden's Ship Tavern. Access to Hayden's shipyard and expertise were of valuable assistance to Bushnell in the development of his dream boat from idea to an actual wood, metal and glass weapon of war.

Phineas Pratt was a member of one of Essex's first families. He was a clockmaker and inventor in his own right. He, in fact, created the machinery and process that turned elephant's tusks into ivory combs that provided Ivoryton with its eponymous industry. It is important to note that, while a young man, Pratt was apprenticed in the ship's carpenter shop of Deacon Abner Parker. The skills he learned there proved to be invaluable, when he assisted Ezra Bushnell in the construction of the hull of the *Turtle*. Pratt's father, Aziriah, was a blacksmith. He would have the skills and equipment to shape and fit the metal pieces necessary to hold the nascent submarine's hull together.

Phineas Pratt enlisted in the Seventh Connecticut Regiment of the Continental army, but David Bushnell requested that Pratt be given leave from his soldiering to assist in the creation of the *Turtle*. This request was granted, and Pratt returned to Essex to become a key member of the top-secret team. Not only did he provide brilliant ideas and technical support, but he also was one of the three truly brave men who volunteered to pilot the machine. They literally went where no man had gone before for extended periods of time—under the waters of the Connecticut River.

Another key co-conspirator was the aforementioned Isaac Doolittle of New Haven. Doolittle, like Pratt, was a clockmaker and an inventor. His many enterprises included a bell work, which gave him access to scarce quantities of iron and steel. Both commodities were hard to come by in colonial America due to British monopolies on their manufacture and distribution. Also in low supply and high demand at the time was gunpowder, the single most important material necessary to wage war. Bushnell, who liked to play with explosives, never had enough on hand, especially when he started to experiment with mines and torpedoes. Much of this problem was solved when Doolittle was granted a license to construct a powder mill in a suburb of New Haven. Bushnell's mine for the *Turtle* required 150 pounds of the stuff, so Doolittle's powder works came in right handy.

Ezra Bushnell, David's brother, was another important team member. Although most people who lived in on the lower Connecticut River were firm believers in the Patriot cause, there were a few Tories sprinkled here and there throughout the area's communities. Each of those Tories was a possible spy who would be more than happy to pass on valuable information to the British, who were firmly in control of Long Island and New York. Brother Ezra, among other reasons, was a vital member of David's team for the simple reason that brother could trust brother implicitly. It was a blessing to David to have someone with whom he could share not only the technical aspects of the project but also the emotional and intellectual stress and strain of the endeavor.

With spies everywhere, Bushnell had to rely on a small network of trusted friends to convey sensitive information about his project. Since the local postmaster was believed to be a British asset, the team resorted to using codes when referring to their clandestine activities. One of their most effective subterfuges was to correspond in Latin. Since that ancient tongue was used primarily by clergymen, whoever was intercepting the mail thought that the team was just arguing the fine points of theology. At one point, the highest-ranking governmental and naval officers among the British occupiers of

New York were apprised of Bushnell's invention and his intention to use it to wreak havoc on the British fleet. Amazingly, Governor Tryon and Vice Admiral Shuldham dismissed this extraordinary intelligence as preposterous. They nonchalantly concerned themselves with which vintage of claret to serve with the beef, and who was that lovely lady at the gala last night?

Secrecy and security were not the only nontechnical challenges Bushnell and his crew faced. Building the world's first submarine proved to be quite an expensive undertaking. The cash-strapped inventor was constantly seeking sources of revenue to provide the wherewithal necessary to create his brainchild. Friends, family and coconspirators were quite forthcoming with money, time, energy and materials, but it was never quite enough to keep the endeavor out of the red.

On February 2, 1776, the indomitable inventor went before the Connecticut Committee of Safety, a body formed by Jonathan Trumbull to administer the state's resources as they pertained to the Revolutionary War. Bushnell requested funding that would enable him to continue the construction of the invention that would ultimately become the *Turtle*. According to the minutes of that meeting, David Bushnell "gave an account of his machine contrived to blow ships & etc." He also swore the members of the committee to complete secrecy in hopes of safeguarding his secrets and designs.

His request was received favorably. The next day, the committee "Moved...D. Gov. Griswold (was directed that) some encouragement should be given to Mr. Bushnell to pay expenses incurred in preparing his machine for the design projected & etc., and to carry forward the plan...it appearing to be a work of great ingenuity...and a prospect that it may be intended with success." Deputy Governor Matthew Griswold personally transported the money from Lebanon, which was the wartime capital of Connecticut, to Saybrook, where the grateful Bushnell was happy to receive it.

Ezra Bushnell was a great help to his brother in many ways. At one point, he purchased David's half of the family farm to ensure that David had the necessary funds to attend Yale College and to conduct his experiments in underwater explosives. Ezra's farm was a secure spot, away from curious eyes, where the first stages of construction of the unique submarine would take place. It was ultimately necessary to move the project nearer to the water, but in the early stages, Ezra's farm was a safe place to start cobbling the creation together.

Ezra's involvement with his brother's invention reached its peak when he was selected to be the first of the three brave souls who would daringly put their bodies in the boat and descend into the depths of the Connecticut

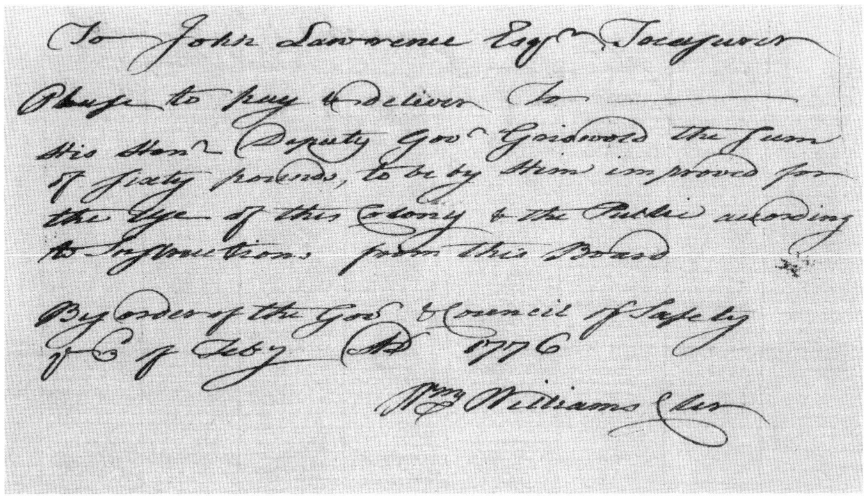

To John Lawrence Esqr Treasurer
Please to pay & deliver To
His Honr Deputy Govr Griswold the sum
of sixty pounds, to be by him improved for
the use of this Colony & the Public according
to Instructions from this Board

By order of the Govr & Council of Safety
ye 6 of Feby AD 1776

Wm Williams Clerk

Connecticut Council of Safety document authorizing Matthew Griswold to give David Bushnell sixty pounds toward the construction of his submarine. *Paul Foundation, Essex, Connecticut; photo by Cultural Preservation Technologies.*

River. This task not only required consummate courage but also called for an extremely high level of physical fitness. The operator had to descend into the depths in darkness and manipulate an array of cranks, levers and foot pedals. These allowed the sub to dive, rise and move through the water. Ezra Bushnell must have undergone months of training to develop the musculature required to perform these complex activities while sitting in a cramped position. A benefit of this training would be that the repetitive exercises, over time, would lead to a slower heart rate, which, in turn, would result in less oxygen required to do the work. This would extend the time that Ezra could stay submerged.

Ezra Bushnell's rigorous training regimen prepared him physically and mentally to perform the tasks necessary to attach a mine to the hull of a British man-of-war. He participated in several test dives off Calves Island in the Connecticut River and prepared himself to undertake the dangerous mission that awaited in the waters of New York Harbor. Alas, poor Ezra was not to fulfill his role in actuating the dreams of his brother. Just before he was to pilot the innovative weapon on its first foray against the Royal Navy's ship of the line, he was taken sick with an illness from which he never fully recovered. It is believed he was stricken with camp fever, contracted during his military service. It was a term that covered typhoid fever, dysentery and sometimes malaria. At any rate, whatever laid him low rendered Ezra

Bushnell hors de combat and forced his brother to procure a pinch-pilot to attempt to complete the mission.

Losing the only trained operator of his invention was a terrible setback for David Bushnell. But he responded to the challenge by contacting Brigadier General Samuel Parsons. Parsons was a native of Lyme, Connecticut, and familiar with Bushnell and his project. He was the commander of the Tenth Continental Regiment and provided Bushnell with three volunteers to train as pilots. Only one of them was to "learn the ways & mystery of this new machine and to make a trial of it," a soldier from East Lyme, Connecticut, named Ezra Lee. Much of what we know of the *Turtle* and its adventures was passed down to us by Ezra Lee. In 1815, Lee was encouraged to write a narrative of his experiences with the submarine by Charles Griswold of Old Lyme. Griswold subsequently interviewed Lee, and his work gives us the most complete record of the attack on the British fleet by the underwater weapon.

Bushnell selected Ezra Lee to replace his ailing brother, because in his words, "he appeared more expert than the rest" of the possible candidates Parsons sent. Within a month of reporting for duty, Lee was undergoing hands-on training under the Connecticut River. The timeframe for his acclimation to learn the operation of what, for the period, was a complicated machine, was shortened by the news that the British occupied Long Island and Washington's army was in retreat. As a result, Lee may have had less than two full weeks to develop and strengthen the muscle groups necessary to run the submarine and to fully acquaint himself with all the intricate aspects of its operation.

The next challenge that that Team *Turtle* need to overcome was the actual transporting of the submarine from its home port on the Connecticut River to New York City, where the enemy had amassed their fleet. The trick was to move a three-thousand-pound barrel one hundred miles in total secrecy. Fortunately, the commercial capabilities of Connecticut merchants provided the logistics to accomplish the task at hand. The *Turtle* was loaded by block and tackle onto a cargo sloop, most likely the aptly named *Crane*, which was built just upriver of Essex in East Haddam, Connecticut. But with the British fleet firmly in control of New York's waterways, it was necessary to offload the *Crane*'s clandestine cargo in New Rochelle. *Turtle* then made its way overland by wagon until it could be launched from land controlled by Washington's army. It was then put aboard another sloop under the command of General Heath. This boat brought *Turtle* to the place it would mount its attack against the British.

As a result, Ensign Ezra Lee found himself, on an early September night in 1776 (the exact date is the object of some speculation), encapsulated in an oddly shaped wooden vessel, being towed down the Hudson (then called the North River) by whaleboats. They were headed in the direction of the British fleet, comprising four hundred ships, as it lay at anchor off Manhattan. The rowers hauled him as close to the big warships as they dared and then cast him adrift. Lee was on his own; his mission was to secure a time-delayed mine to the hull of the HMS *Eagle* and then get the hell out of there as best he could.

Unfortunately for the brave Patriot, his team had badly misjudged the tides. He was swept past his target and was forced to spend two and a half grueling hours fighting against a strong current with only one arm turning the sub's small, poorly designed propeller. This was a terrible state of affairs for Lee. The *Turtle* was not designed to function in strong currents. Its propulsion system was just not up to the task. It is believed that Lee was able to maneuver his unwieldy charge into the lee of Bedlow's Island (now the home of the Statue of Liberty). This would have allowed him a bit of respite after the arduous work of working the cranks and pedals.

Finally, the tide slacked enough to make it possible for the now somewhat rested submariner to launch his assault on HMS *Eagle*. By this time, he had a new problem with which to contend. With each passing minute, the night was hurtling toward the coming dawn. If the *Turtle* was spotted by a British sentry, the veil of secrecy would lift and Lee would be the proverbial sitting duck. He finally reached his quarry before the dawn gave him away. He was close enough to hear the voices of the English sailors and deck. In his later account, he claimed that he was able to actually see some seamen on the deck of the huge, British warship. If that was the case, he would have been vulnerable to musket fire should one of the enemy spot him bobbing against the warship's hull.

In any event, Lee, in his words, "shut down all the doors, sunk down and came up under the hull of the ship." This was the moment for which the *Turtle* was designed and built. This was the very first attack on an enemy vessel by a submarine. Lee was now in almost complete darkness. His compass and depth gauge were ingeniously illuminated by foxfire, a phosphorescent fungus that grows on the north side of oak trees, but for all intents and purposes, he was diving blind. He used his foot pedal to let water into the vessel so he could sink under the surface. When he believed he was at the correct depth and the right position under the hull of his prey, Lee was ready to attach a deadly mine, full of 150 pounds of gunpowder with a timing mechanism, to the hull of the HMS *Eagle*.

This was the critical moment. All of the inventiveness, planning, training and hope that drove the *Turtle*'s mission was about to come to fruition. The bomb would be attached to the *Eagle*, the *Turtle* would scuttle away and *boom*—the pride of the Royal Navy would be blown to smithereens. Except, things didn't turn out that way. The brave submariner was not able to get the wood screw built into his submarine to penetrate the hull of the British ship. He tried another spot, but to no avail. The drill just would not work. His capsule was rapidly running out of air, daylight was looming, Lee was exhausted and the only option was to surface and try to get back safely to the protection of colonial cannons. The one-of-a-kind weapon failed to successfully complete its first foray against the enemy, but it made a whole lot of naval history in the process.

Lee made good his escape, but it was a close thing. Since dawn was brightening the horizon, he no longer enjoyed the protection of invisibility. He needed to stay near the surface in order to breathe and to see where he was headed. At one point, he was spotted by British troops stationed on Governor's Island. They immediately manned a twelve-oared barge and gave chase to the dog-tired submariner, who was still gamely cranking away to turn his two-bladed propeller. As the red-coated rowers came within fifty yards of the retreating hero, Lee decided he would cut loose his explosive device in hopes that the enemy would stop to retrieve it. His intention was to blow them up, but he also realized he was likely to be hoisted on his own petard with this maneuver. Rather than pick up the mine, the British turned tail and scurried back to the safety of their fort. Lee was overjoyed at this turn of events, in that he lived to tell the tale.

When the exhausted ensign jettisoned his explosive, its timing device was automatically set. One of the many positive things that came out of this initial attempt at underwater warfare was the fact that the timing device that Bushnell had devised worked perfectly. The fleeing Lee now had a fair tide drawing him away from the enemy and toward the safe haven of the colonial troops. He came into view of the Americans, and they sent out a whaleboat that took the tired and sore sailor in tow and brought him ashore safe and sound. The spectacular explosion was an eye-opener for both the Royal Navy and the colonial forces who saw and heard it and assessed its destructive potential.

The *Turtle* was involved in two more attempts to inflict damage on British ships on the Hudson, but neither was successful. Although Bushnell's early efforts came to no avail, the sheer inventiveness of his creation set the stage for a whole new dimension in shipbuilding and naval warfare. Phineas Pratt never received a pension for his service to the Revolution. His involvement with the *Turtle*, however, proved of lasting benefit to the citizenry of Essex. He fashioned some of the metal of the submarine into the Essex town clock, yet another timing device that grew out of Bushnell's work. Ezra Lee was also unsuccessful in trying to get a pension from the government. He claimed that he was stricken with debilitating rheumatism, as a result of sitting in cold water up to his knees while piloting the first undersea warship.

David Bushnell continued to develop underwater explosive devices after his work with the *Turtle*. He met with the Connecticut Council of Safety in New London, a trip that included a visit to the Essex-built *Oliver Cromwell*, as well as Fort Trumbull and Fort Griswold. He collaborated with Governor Jonathan Trumbull, the Patriot whom Washington called "Brother Jonathan." The two men decided to focus on floating and submersible mines. Bushnell attempted to blow up HMS *Cerberus* in New London Harbor with a mine. The mine was spotted by some sailors on a nearby schooner, who dragged the bomb onto their boat, only to have it explode with fatal results to three sailors.

Bushnell then began to experiment with a simple design based on a gunpowder keg with a flintlock mechanism to detonate it. He planted several of these devices in the Delaware River in hopes that they would drift down on the British fleet anchored off Philadelphia. The river was full of ice, which slowed the progress of the mines. One was fished out of the river by two boys, who quickly were rewarded with a quick trip to eternity for their curiosity. This explosion alerted the Royal Navy to the mines, and their sharpshooters soon picked off the drifting barrels one by one, until the danger passed. The incident was later immortalized in a poem called "The Battle of the Kegs."

Perhaps burned out by his repeated failures, David Bushnell decided to return to Yale and pursue a master's degree. But the war was still raging, and he was caught in a raid by British troops and taken prisoner. He was in a very tight spot. If his captors were to find out about his experiments with explosives and his attacks on the Royal Navy, things would not have gone well for the scholar/inventor. The specter of confinement in one of the truly hellish English prison ships that held those unfortunate enough to be captured by the enemy was a terrifying prospect. For all the British knew, he was a civilian, but with their espionage system he could be identified as a Patriot warrior any minute. Fortunately, Brother Jonathan Trumbull came to Bushnell's aid and arranged to have him repatriated as part of a prisoner exchange. This was a life-saving circumstance for Bushnell. His sigh of relief must have been a long and happy one.

Given a second chance, Bushnell once again took up arms against the Crown. He was commissioned a lieutenant in an engineering unit of sappers and miners. He was promoted to captain and, apparently, was not looked too highly on by the soldiers under his command. On one occasion they attempted to fill his canteen with gunpowder to blow it up. This was a scheme that could have brought a quick end to Bushnell's military career. The disgruntled soldiers were dissuaded by one of their compatriots and eventually gave up the plot. Bushnell served out his commission until the end of the Revolutionary War.

But Bushnell, like Pratt and Lee, was not satisfied with the compensation that he received for his service. He was given a land grant and a sum of money, but he did not think that he was treated fairly. He left Connecticut and became somewhat of a man of mystery. There were many rumors about what happened to him. Had he fled to Europe? Did he become a drug addict? An alcoholic? Had he gone insane? Was he an invalid? A suicide? The answer was much more prosaic. While he did travel to France, he returned to the newly minted United States. The complex man became a professor of medicine at a small college in Georgia, using the name David Bush.

According to Fred Frese, one of Bushnell's daughters had a romantic relationship with Robert Fulton. Fulton is generally acknowledged as the inventor of the steamboat, when in reality, he basically stole the ideas of Connecticut River inventors Samuel Morey and John Fitch and used his superior business sense and marketing skills to bring the world-changing types of vessels into common use. The interesting connection with Bushnell and his ideas focused on Fulton's later experiments with timed underwater explosive devices, submarines and propellers. It is an intriguing thought experiment to speculate on the connection between the ideas of the two men, both of whom left their imprints on the maritime history of the world.

The aforementioned Fred Frese is the dynamic force of nature who took a taproom conversation and transformed it into three replicas of the *Turtle* that may be seen and admired today. He is also the coauthor, along with Roy Manstan, of the definitive book that covers all aspects of the submarine's history, technical specifications and recreations. *Turtle: David Bushnell's Revolutionary Vessel* is a monumental piece of research and writing. Frese's motto is, "If you are not living on the edge, you are taking up too much space." He is a retired technical arts teacher at Old Saybrook High School. He has inspired generations of students to build human-powered submarines and to develop and harness their personal energy in productive channels throughout their lives. Roy Manstan is a mechanical engineer and former command diving officer of the Navy Undersea Warfare Center. Together they form a team that brought this fascinating chapter of Connecticut's maritime history to life to be shared and enjoyed well into the future.

Fred Frese decided to create a replica of the *Turtle* to celebrate the bicentennial of the United States in 1976. He collaborated with a writer named Joseph Leary, and they named their re-creation *American Turtle.* The chairman of the board at Electric Boat was very supportive of the project. The vessel was launched from Steamboat Dock in Essex in August 1977. Bill Winterer, who owned the Griswold Inn, the venerable watering hole in Essex, arranged for the popular Governor Ella Grasso to smash the requisite bottle of champagne against the hull, and into the Connecticut River *American Turtle* went, following its predecessor who made the same splash some two hundred years prior. You can visit that replica today at the Connecticut River Museum at the foot of Main Street in Essex. You can turn the cranks and press the foot pedal, just like Ezra Lee did on his foray against the British.

Fred was once again inspired to re-create the *Turtle.* He teamed up with Roy Manstan, like Bushnell a Westbrook native. This inspiration blossomed into the Old Saybrook High School *Turtle* Project. According the Frese and Manstan, the goal was to build a replica "capable of being launched and operated in actual at-sea conditions." Their hope was to re-create as closely as possible the original and to test it in terms of "propulsion, transiting and maneuvering." They did have an overriding concern for the safety of the operators and included things not available to Bushnell such as O-rings, modern valves and pumps and Lexan instead of glass for the windows and gauges.

The project proved to be a splendid opportunity for the students of Old Saybrook High School. It was marked by inclusivity; both college-track students and those with technical inclinations were welcome. According to Frese, "the only criteria were motivation and creativity." Beginning with the initial stages of construction in 2003 to completion in 2007, "dozens (of students) contributed their energy to the *Turtle* Project." They cut planking, welded and contrived mechanical systems. Community volunteers were also instrumental. They provided classroom support, research, logistics and moral support. Some students came back after graduation to see the project through to completion. The project was truly a community effort that provided students with knowledge, self-confidence, technical skills, teamwork and an experience of a lifetime.

It was decided by the team that, for safety's sake, the pilots of the new boat should be U.S. Navy–trained divers. The Navy Undersea Warfare Center developed an educational relationship with the high school students, which was a unique opportunity for the budding scholars to expand their intellectual horizons. The navy also agreed to provide divers for initial trials, the christening and launch and later sea trials. A team of Engineering and Diving Support Unit were the pool from which the project's test pilots were selected. The three selected were Roy Manstan, Paul Mileski and David Hart. Five navy divers would be handy-by whenever the *Turtle* was in the water.

The boat made contact with the water of the Connecticut River on October 22, 2007. This splash occurred at the Essex Boat Works, a venue with its own historical Essex significance. The boat works' crane lifted the replica off the hard (as sailors call dry land) and deposited it into the waters of Middle Cove. The newest *Turtle* was maneuvered into a slip, and Roy Manstan checked out the boat in terms of safety and mechanics. He pronounced it good to go. While it was at the dock, it sat in the water for a few days to swell up so that there would be minimal, if any, leakage.

Wick Griswold and the University of Hartford students in his Sociology of the Connecticut River class were at the steamboat dock in Essex on a very chilly, windy Saturday morning on November 10, 2007. We witnessed the christening of the latest iteration of the *Turtle*. State senator Eileen Daly did the bubbly bottle honors, and the egg-shaped boat was lifted of the dock in front of the Connecticut River Museum and plunked into the cold river water. Roy Manstan, snugly stuffed into the capsule, put the project's product through its paces. It performed admirably and returned to the dock, its Revolutionary War–era Don't Tread On Me flag snapping smartly from a staff attached to the hatch. The launch was declared a success.

The replica then was transported by truck to Mystic Seaport, where the museum's shiplift apparatus made it possible to do a series of operational tests and experiments. It was evaluated for a variety of performance-related processes such as propulsion, diving, submerged handling, physical effects on the pilot, instrumentation and many other systems and maneuvers. The results provided a tremendous amount of technical data that yielded tremendous insights into Bushnell's remarkable invention and all that it entails. In the words of Frese and Manstan, the project "put the finishing touches on gaining a better understanding of the revolutionary thought that led Bushnell and his collaborators to put their lives as risk in the pursuit of American independence." The boat can now be seen at the Nautilus Submarine Museum in Groton, Connecticut.

But the indefatigable Fred Frese was not done with human-powered submarines. He marshaled a new cadre of Old Saybrook High School students and led then into a new chapter of underwater adventures. Frese and his students designed and built a series of single-person submersibles that were propelled by students in scuba gear pumping madly away at pedals while lying horizontally in a slim, fiberglass, hydrodynamic tube. The team worked with engineers at Electric Boat to design propulsion systems and other aspects of the boat. The students all had to be SCUBA certified to fifty feet in order to pilot the craft.

For a number of years, OSHS teams would pile into a van with their submarine on top and drive to Maryland to compete in an International Engineering competition featuring human-powered submersibles. There boats were always named *Miss Jesse* in honor of Frese's wife. The races were held at the Navy Model Basin, in a specially designed forty-foot pool. There were only two high schools that would routinely compete against the likes of teams from MIT and Rolls-Royce. Fred's teams would at least hold their own and sometimes won! Of the students who participated in the program over the years, at least twenty went on to become engineers in their adult lives. They often come back and get in touch with Fred Frese to thank him for all he did for them. For the dedicated teacher, that is the real reward of the job.

Growth and Prosperity

Before the American Revolution, few people resided in Potapaug Point, but during and after the war, significant cultural and demographic changes took place. New ships were needed, not only to secure harbors and coves or stave off British forces, but also to maintain and develop more trade. Shipbuilders were drawn to the area because the natural environment afforded deep sheltered coves, and desirable trees were plentiful. By the end of the eighteenth century, new shipyards were established on the western shoreline of the two lower coves and the larger North Cove, then referred to as "Falls River Cove."

The Falls River bisects the town from east to west. Dams built in Centerbrook, the original village center of Potapaug in the early 1700s, powered several types of mills, including a gristmill, a sawmill, a trip-hammer shop and ironworks, all of which were within one hundred yards of the church. In 1722, everyone was a member of the Puritan Church, and the Connecticut Colonial Court agreed to the establishment of another ecclesiastical society at Centerbrook, as it was determined that there were enough residents in the village to support the building of a second church. Prior to that time, residents had to travel rough roads and waters in all types of weather to get to church in Saybrook. The decision to build a second church brought a communal sigh of relief.

A Congregational church was erected in 1724; it was renovated in 1757 and rebuilt in 1780 to accommodate a larger parish. Farming and the church dominated the local culture until the American Revolution, but after the war ended in September 1783, Centerbrook was no longer Potapaug's center. Village activity shifted to the Point. This was due mainly to the increased demand for ships, the rise in shipbuilding on the coves and an influx of skilled workers and their families from diverse backgrounds moving to Potapaug Point to work in the shipyards.

The Haydens' yard attracted many people to Potapaug Point, but despite their wealth and prominence, like other families, the Haydens experienced times of travail and sorrow. In the spring of 1791, Uriah and Ann's son Captain Nehemiah Hayden sailed south along the Eastern Seaboard on a trading venture. News traveled slowly in those days, but eventually word reached Essex that thirty-five-year-old Nehemiah died on May 29 of that year. We can imagine the grief experienced by Nehemiah's wife, Sarah, their young sons, his parents and the entire extended Hayden Shipyard family. Many lives were lost at sea and in foreign lands, and it was often months before word of an accident or loss of a loved one reached their families. The inscription on Nehemiah's grave at Riverview Cemetery reads, "Died in St. Georges on the Island Bermudas."

The Hayden family shipping business greatly influenced the shifting of the village center from Centerbrook to Potapaug Point, as well as the shift from an agricultural economy to one based on the developing shipbuilding industry. Construction of vessels of all types continued in a pre-industrial manner. Each vessel had its own unique design, purpose and individual characteristics based on the builder and/or purchaser's desires and finances. If ships of the late eighteenth century were all lined up, stem to stern, no two vessels would look alike, for each was built to individual specifications.

The new nation required ships for both domestic and international trade. Between the American Revolution and the Civil War, five hundred to six hundred ships were built in Essex. Several stores were built to accommodate the needs of the shipping industry, and most mercantile businesses were situated at the lower end of Main Street near the shipyards. Ebenezer Hayden had a general store constructed that was vital to the community and located near the Pratt Smithy in Champlin Square.

Another result of the influx of people who moved to the village and of folks who improved their financial status toward the end of the eighteenth was the increase in houses built in the area. In 1776, there were six houses either under construction or completed on Main Street. By 1810, there were thirty homes in the area.

To accommodate the increasing requirements of the shipyards and individual shipbuilders, a ropewalk was constructed in 1797 by Grover

L'Hommidieu and his partners, Ebenezer Hayden's son Jared and Sala and Reuben Post. The ropewalk was 15 feet wide and 990 feet long and ran parallel to the north side of the present Main Street. This was the first cordage company in Essex, and various sizes of tarred ropes were made there for different purposes. Some were required for standing and running rigging and other critical ship needs. These included thick hawsers, sturdy ropes used for securing large vessels to moorings or other heavy hauling tasks, like when a ship was stuck on the ways.

The Hayden Shipyard was enhanced by the addition of master carpenter Richard N. Powers, who resided on Uriah Hayden's property and married his daughter Abby Ann on December 14, 1826. The five-acre yard's reputation for shipbuilding and shipping grew steadily. Uriah's wharf buzzed with activity, as did his shipyard, and his family grew and prospered. But Uriah was not the only Hayden who was in his prime and flourishing at that time.

Uriah's cousin Ebenezer Hayden mulled over key aspects of a large development initiative he imagined for an undeveloped section of Potapaug.

Oil painting, interior of the ropewalk. *Courtesy of the Essex Public Library; photo by Shirley Malcarne.*

While the desire to increase his fortune played a part in his elaborate plans, Ebenezer was a visionary who had the financial means to make his dreams a reality. He was equally as talented and ambitious as Uriah. The Hayden cousins were two of the most influential and prominent merchants of Potapaug.

After his wife, Prudence, died at the young age of thirty-five in 1779, Ebenezer Hayden continued to live in his home on West Avenue and bring up their children. While Uriah was occupied with his shipping and shipbuilding enterprise and the establishment and construction of the Episcopal church on Pound Hill in 1800, Ebenezer Hayden continued to improve his general store in the shopping center of town. The establishment provided food items, local produce, home goods and popular items such as soap, cigars and candles for the general population.

Ebenezer took action on his innovative plans to develop an area on the west side of North Cove, near Denison Point. His plan required waterfront property, and in 1786, he purchased thirteen acres of land at Denison Point on the North Cove from Jabez Denison. Six years later, in 1792, Ebenezer purchased another thirty-three acres from Denison property owner Robert Denison, younger brother of Jabez. The total Denison acreage purchased from the brothers was forty-six acres. The property ran from the base of the North Cove to Denison Road in Meadow Woods. These key purchases of land allowed Ebenezer to move forward and realize his objectives.

In retrospective, it is obvious that Ebenezer Hayden envisioned far more than a quaint and separate housing development or neighborhood. Ebenezer had a vivid imagination and the intelligence, fortitude and financial means to develop every aspect of his complex plan. He began work on what became known as New City, a place that would attract the resources of shipbuilders and their families, shipmasters and shipping merchants seeking ever-larger ships and those who were interested in developing the shipping industry in Potapaug.

Ebenezer Hayden solicited help from friends. He was well acquainted with Benjamin Williams, who set up his forge on the Falls River about 1761 and lived in the area for over thirty years. In the summer of 1792, Benjamin Williams bid on an old building with an eye to help Ebenezer Hayden set up a shipyard at New City on the banks of North Cove.

In 1789, after several repairs to the 1724 meetinghouse of the 2nd Ecclesiastical Society and much controversy over the building's future, the community finally came to the consensus that a new meetinghouse should be erected on land a few rods west of the old house. In August 1792, the building was put up for auction, and Benjamin Williams came in with the highest bid, twenty-five pounds. Williams was able to purchase the thirty-by-forty-foot structure, which soon became an essential storehouse and workshop at Ebenezer Hayden's New City Wharf. The old meetinghouse building remained in the New City Shipyard from 1792 until it was torn down in 1860. While there is no evidence to support an official partnership between Ebenezer Hayden and Benjamin Williams at that time, the act of dismantling and moving a sizeable building down to the North Cove wharf documents the fact that both Ebenezer and Benjamin were working toward the same goal of setting up a shipyard at New City.

The Hayden and Williams commercial relationship endured through at least four generations and into the last half of the nineteenth century. Benjamin Williams son Samuel moved to the Falls River section of Potapaug as a young man to run the small gristmill next to his father's forge. He arrived there in 1777 and soon built a sawmill across the Falls River from his mill. In order to accommodate the needs of his growing family, Samuel moved into a house on the east side of what is now North Main Street. Robert Denison built the house in 1767 and mortgaged his family homestead to Samuel Williams in 1792.

Samuel continued to enlarge his business complex on the Falls River as his and wife Irena's family grew. By the time his sixth son, Richard Pratt Williams, arrived in 1796, Samuel was about forty-five years old. He had a new baby, two daughters and five older sons and a shipyard that he built in 1794 on the north side of his Falls River commercial complex. Like his father, Benjamin, Samuel was an enterprising man who did not hesitate to build or repair dams with his young sons or build schooners and sloops in his shipyard while overseeing the sawmill, lumberyard, gristmill and forge. In time, Samuel Williams owned over sixty acres of land on both the north and south sides of the Falls River.

The Williams Shipyard became an important addition in Samuel's business section of Potapaug, known as Meadow Woods. The commercial district was located on both the north and south sides of the Falls River. Between 1989 and 1991, Wesleyan University students working with a professor, Dr. John Pfeifer, and Essex town historian Don Malcarne conducted archaeological research of the Williams complex. Their work revealed that the Williamses' vertical business area included not only their grist- and sawmills and a lumberyard, but also the shipyard was established with a yard for constructing vessels, areas for making ship repairs, a slipway for launching, a bulkhead and a large building, possibly a workshop and woodshed. Down the river to the east, not far away on the southern shore,

Williams Sawmill and Dam in Meadow Woods, Essex, Connecticut. *Courtesy of the Essex Historical Society.*

was a water fence where lumber was corralled and kept wet in a secured area until needed. The Essex Historical Society and the Essex Land Trust booklet *Follow the Falls* reveals that Williams prepared and sold complete packages for building both houses and ships. Each prefabricated package contained everything needed to complete the project, including the nails.

In time, Samuel Williams's son Ezra added a comb and ivory cutting business to the Williamses' commercial enterprises on the Falls River. As Samuel's sons grew up and learned various trades from their father and the many skilled artisans who worked for him, they were able to take over the daily operation of the forge, mills, shipyard and business transactions. The Williams family also had interests in other ventures besides their Meadow Woods businesses, including agriculture and a fishery near the family's shipyard. Both of these concerns were valued about the same. Diversification of products and services was advantageous to the extended Williams family.

In 1798, Ebenezer Hayden moved to a house on Main Street with a storied past connected to its construction. Oral tradition reveals that roof timbers were made in Windsor, Connecticut, and purportedly shipped downriver where they were hauled up Main Street, probably by ox train, and assembled at Hayden's new house site.

As early as 1795, Reuben Post and his brothers Sala and Levi, house builders, shipbuilders, shipmasters and part owners of the first ropewalk, were also partners in Ebenezer's shipping business. They leased land from Ebenezer and constructed their homes on three of his lots. Their brother Captain Ward Post joined them a few years later. After their houses were built, the first Post residents of New City purchased the land on which their houses stood from New City developer Ebenezer Hayden. In 1797, Reuben paid eighty dollars for his roughly one-quarter-acre plot. Paying for land after building a house on it was a common practice in the area for decades.

The Post brothers of New City were four of thirteen children of miller and deacon David Post and his wife, Deborah Ward Post, who purchased a newly constructed home on Denison Road. In all, the Posts had nine sons and four daughters, ultimately resulting in 102 births of their children's offspring. Thirty-two years after moving into the house, Deacon David Post purchased the property his home was built on. David Post operated a gristmill in the 1760s on the Falls River, just to the east of where the Williamses later built their mill.

Ebenezer Hayden's affordable housing development for shipmasters and shipbuilders was one main consideration in his grand scheme for New City. Another priority was the design of the dock. New City dock ran out into the cove from the shore to the edge of the channel, and its orientation into the cove, angled slightly toward the launch area on land, was particularly advantageous to both mariners and shipbuilders. With the New City wharf extending out to the channel, hulls launched from the ways on shore could be easily moored at the end of the dock. Those that were finished off at New City could enter the channel immediately and head off to New York and other cities.

The depth of water in the coves around 1800 was considerably deeper than it is today, and vessels were able to navigate the coves in ways that are not possible today due to the amount of silt that has built up over time. This gradual lessening of the coves' depth occurred mostly as a result of floods and the destruction of beaver dams that caused silt to flow down the rivers and into the coves.

Much consideration and planning went into the design, layout and precise positioning of all elements in Ebenezer Hayden's New City. This included its highway, shipyard, landing, extended dock and store and the

Map of Essex Shipyards. *Courtesy of R. Major.*

surrounding neighborhood. As the eighteenth century came to a close, Ebenezer Hayden was not only the leading mortgage holder in the lower Connecticut River Valley but also a major financier of ships. He positioned himself to continue to increase his fortune in the new century with further development of his New City neighborhood and shipyard enterprise.

Shifting Tides

When the nineteenth century began, the stage was set for continued commercial and economic growth at Potapaug Point. According to J.H. Beers, "During the first years of the century, 1,200 to 2,000 tons of shipping were built" there per year, mostly for use in the coastwise and West Indies trade. A period of great activity and progress provided residents with a number of reasons to feel optimistic. More bottoms were built, and new shipyards emerged to meet the demand for vessels to supply eagerly sought-after trade goods such as sugar and rum.

Several shipbuilders were active in the early 1800s, including the Pratt brothers, who worked adjacent to the New City yard. Uriah Hayden was nearly seventy years old in 1800 and no longer building ships himself, but he made certain his shipyard was in competent hands for the future. He sold his yard to his nephew Calvin Hayden, a shipwright who crafted vessels at Hayden yard for the next thirty years. Amassa Hayden also started in the business about the same time and continued for three decades. John G. Hayden, older brother of Amassa and Richard Hayden, built his home in 1800 and was also a shipbuilder and master figurehead carver.

Ancillary shops were established in order to support the growing shipbuilding industry. In January 1800, Samuel Lay, descendant of first settler Robert Lay, leased a shop to Ezra L'Hommedieu at the foot of Main Street near the wharf to set up a carving business. Spar maker Gamaliel Conklin, who purchased Ebenezer Hayden's West Avenue home in 1798, built his own house on Main Street between 1800 and

1802. Gamaliel owned a block and spar shop with his partner Jesse Murray, and the two made masts, spars and running gear for the local shipbuilders and masters.

Ebenezer Hayden worked on his New City development project. In 1803, he cleared a new road, called a highway, along the south side of his New City property. For several years, the new road was simply referred to as "the road to the shipyard." The road spanned a quarter of a mile, west to east. It was thirty-three feet wide and passed by the Post brothers' houses under construction, all facing Ebenezer's road on the north side only, keeping the houses within the bounds of his property. The road continued down toward the North Cove to Pratt's "mear stone," a stone that delineated a corner of Judea Pratt's property located adjacent to the shipyard and cove. At that point, the road narrowed into a cartway, called a "strate." It paralleled the edge of the cove and ended conveniently at the base of Ebenezer Hayden's dock, wharf and landing. New City Shipyard was the destination of most of the early inhabitants who used the new road.

The home lots on the north side of New City Street were 165 feet deep and varying widths. The last property on the road nearest the cove was intriguing to historian Don Malcarne. The Pratt property was located next to a small parcel with a blacksmith's shop beside the New City Landing. Malcarne informs us, "The Judea Pratt property to the west and this parcel were one and the same for a long period. When Judea Pratt picked up the land where his blacksmith shop stood (adjacent to the landing) is somewhat of a mystery." Judea purchased the land on which he built his home from Ebenezer Hayden in 1803 and did not sell his home until 1840. His adjacent blacksmith property was not sold until 1841. Whenever the blacksmith shop was set up, its proximity to the cove and to New City wharf was advantageous to shipbuilders who built their vessels on or near the western shore of the cove as well as to the Pratt brothers. Judea and his brother Asahel built vessels there on the south side of the landing for the first decade of the nineteenth century.

Narrow cart paths in Potapaug widened with more frequent use and larger wagons. New roads were laid out. In an effort to provide public access to his development and shipyard, Ebenezer conveyed his new highway to the community in May 1803, but it was thirty years before the road would be officially named New City Street.

Based on the need for more housing and rooms, another tavern opened on Main Street. In 1806, Ethan Bushnell bought a home from Richard Hayden, who had built the house five years earlier in an effort to expand the innkeeping business and provide more lodging for the town. The property came with a barn, and Bushnell turned the house into a tavern, known then as Ethan Bushnell's Tavern. His son Sidney became a part owner in the business, and he and his sister Lucy inherited the property when Ethan died in 1849. The tavern changed hands several times before becoming the Griswold House, as it was called in 1865, according to an old postcard of the place. As current owner Geoffrey Paul said, "Born of Independence, these are the beginnings of the Griswold Inn, a destination for seafarers, wash-a-shores and travelers alike for nearly 250 years."

Today, the "Gris" remains the center of a thriving community, offering a unique experience and a comfortable connection to the past. The beloved, renovated and greatly expanded Griswold Inn houses one of the most extensive collections of historically significant works of marine art in the country. Owner Geoffrey Paul is an expert on the Gris's museum-quality maritime art and historical collections. He offers guided tours free of charge, though reservations are required.

November 24, 1808, was a day the Hayden extended family, including their shipyard workers, had etched in their minds for the rest of their lives. Esteemed Captain Uriah Hayden died that day, and no doubt most of the town turned out for his funeral. Essex still recognizes Captain Uriah Hayden's contributions, through paintings, history lessons and wonderful walking tours led by the energetic and knowledgeable docents from the Essex Historical Society.

In his will, Uriah left an undeveloped shipyard parcel to his grandsons Uriah and Richard. Perhaps he hoped to ensure that the family shipbuilding enterprise would have room to expand through their efforts. In 1809, three of Uriah's sons went into business together. Samuel and Ebenezer Hayden opened up a store in a brick building on Main Street. The store was a chandlery, or ship's store and warehouse, and the Haydens' brother-in-law Timothy Starkey joined the business the following year. The establishment was called the Hayden Starkey Store. At about the same time, Ezra L'Hommedieu invented the double twist auger, a tool that was a great boon to home and shipbuilders, as it allowed them to bore clean holes much faster in the thick hulls of ships or frames of houses and barns. All over town, the building boom continued.

Ebenezer Hayden worked independently on the road and housing components of his New City, but the shipyard was developed in collaboration with others. New City's shipyard appears to have been created in tandem with the Williamses from the earliest of days in 1792, when Benjamin

Williams hauled the old meetinghouse building to "Ebenezer Hayden's Wharf," as it is named on an 1803 survey map of the New City Shipyard. The wharf description on this map reveals that Ebenezer reserved his right, and that of his heirs, to "wharfing at the channel," and it stated that the area was "containing by estimation one acre and half be the same or less." Immediately to the north of the wharf, along the western shore of North Cove, a rectangular plot is defined with the title "Ebenezer Hayden's Shipyard Property" written within. No one except Ebenezer Hayden is associated with the ownership of the New City Road, the narrow strate to the wharf landing, or the shipyard, in 1803.

In an unpublished article titled "The Founding of New City," former Essex town historian Don Malcarne wrote:

> *A contemporary business associate of Ebenezer Hayden was Samuel Williams....Mr. Williams and his sons worked in conjunction with Ebenezer Hayden on the establishment of the New City wharf and shipyard—it is presumed, based on archival and archaeological study, that vessels built in the Falls River "yard" were "fitted out" at the New City.*

By 1805, the economic tide began to shift as Britain and France became embroiled in an extended period of conflict. News of the hostilities reached Potapaug though mariners returning from European ports, and for a time, most New Englanders were relieved that America remained neutral and was not drawn into the conflict. As time went on, both French and British warships seized American vessels. The British also claimed the right to board American ships to reclaim British citizens, and in the process, Americans were illegally seized and forced into service in the British Royal Navy. By the end of 1807, President Thomas Jefferson, with the approval of Congress, tired of affronts from both Britain and France. He signed the Embargo Act on December 22, 1807, prohibiting trade in all foreign ports.

News of the embargo hit the shipyards like a cannon ball hitting a man-of-war broadside. Shipbuilding and commerce were devastated—not just in Potapaug, but all along the Connecticut River and up and down the Eastern Seaboard. United States exports fell 75 percent in just one year. President Jefferson probably had good intentions, but the embargo did not alter British policy, despite it being in effect for fourteen months.

While shipbuilding and merchant trade were at a standstill and some Federalists cried out for secession, most folks in Potapaug and surrounding communities were as they are now, proud Yankees who wanted to return to work at the docks and yards and reestablish trade. As soon as the embargo ended, progress went full bore ahead again.

Beginning in 1810 and for the next ten years, more of the original Lay property was sold off by Robert Lay Jr.'s son Samuel. George Harrington, Felix and Tim Starkey and Asahel Pratt bought eight acres between them and began building houses on what was then called New Street. In 1813, Richard Hayden opened a big ship chandlery and general store on the corner of Novelty Lane. It remained there for nearly half a century before being moved to its present site on the south side of lower Main Street in 1849.

Construction of a new, longer ropewalk was also initiated in 1813. It ran parallel to the north side of Pratt Street, about 250 feet north of the earlier ropewalk. Unlike the first, open-sided ropewalk, the new ropewalk had a covered roof and sides, an office at one end and an enclosed tar house. The structure was 1,200 feet long and 22 feet wide, allowing for greater lengths of tarred rope to be manufactured.

Perhaps its most unique feature, by today's standards, was the means by which the mechanical components of the ropewalk were powered. A workhorse spent its days tethered to a treadmill bar in the basement of the ropewalk. The animal's sole purpose was to walk round and around in circles all day to power the ropewalk machinery twisting ropes in the ropewalk above. Gurdon Smith and Sala, Reuben and Levi Post were the first proprietors, so they must have transferred their ropemaking and marketing skills to the new establishment. For about eighty-five years, until the ropewalk was torn down in 1898, the longer, second ropewalk had a number of owners and produced miles of cordage. In later years, when the need for a cordage company dwindled, fish line and nets were made there.

Beginning in June 1812, the United States was at war with Great Britain over the impressment of seamen and access to foreign markets. The Royal Navy established a blockade of the Connecticut River and Long Island Sound. Essex-based vessels made attempts to slip past, but many were captured. Small boats from the river and New London would try to blow up the warships with mines and bombs, with little success. One sailor, Torpedo Jack, was caught while trying to row to the safety of the Connecticut River. He may have been forced to provide intelligence to his captors about the navigability of the river and the layout and defenses of Essex Harbor.

Shipbuilding and commerce up and down the Connecticut River were compromised by the blockade. Exports were down 75 percent. The number of ships being built decreased. Connecticut ship merchants such as Richard Hayden, then owner of the Hayden Shipyard, faced severe financial

View of the second Essex ropewalk. Engraving by John Warner Barber, 1836. *Courtesy of the Connecticut River Museum.*

losses. In an effort to recoup a fraction of his losses, Richard Hayden converted some of his ships to privateers and sent them out to confiscate British ships and the valuable cargoes they carried. Some villagers said the Haydens were responsible for drawing attention to Potapaug, as they advertised in a New York newspaper the sale of a new 301-ton schooner and revealed that the vessel could be fitted out as a privateer. Others believed harassment by Essex vessels of the British ships blockading the Long Island Sound was to blame for the British targeting their ships. Whatever, the case, Essex was about to feel the full weight of the British wrath.

The British Raid and the Rise of New City

Through the grapevine of sympathizers and efforts of spies, the British got wind of the fact that at least four other privateers were being built in Potapaug, all with equally devastating potential as the vessel advertised in New York. The British navy reacted fast and furiously to American interference at their blockades of New London and the sound and the threat posed by Essex-built privateers.

During the night of April 13, 1814, while most of the townsfolk slept and fog rolled over newly plowed fields, a catastrophe loomed. British commander Captain Richard Coote was wide awake. To crews from several British warships, Coote gave orders for six ship's boats filled with marines, soldiers, arms and supplies to head to Saybrook Bar. At the mouth of the Connecticut River, the commander dispatched the boats upriver to Potapaug. Captain Coote was probably well acquainted with the area. Purportedly, disguised as an old clam peddler, he had surveyed the village. While this story may have originated in an old salt's yarn, Captain Coote's scouts surveyed the vessels in the coves and Falls River. They knew precisely which vessels they would target.

Armed with torches and swivel guns, the British proceeded up the Connecticut River. They were almost thwarted in their efforts to reach Potapaug Point, not by local militia but by strong northwesterly winds. They did not reach the Point until about 3:30 in the morning.

The local militia were at the ready, but the defenders were far outnumbered and no match for the mighty British navy. A brief skirmish left the locals unharmed, but the British made it crystal clear. If folks wanted their lives and properties to be spared, they must not resist. Captain Coote ordered his troops to carry on with their mission of arson and pillage. After gathering in squadrons, some of the marines rushed off to confiscate supplies from the taverns, chandlery and stores. Others headed to the piers, shipyards and coves to carry out their destructive deeds.

Flames and smoke from burning ships aroused more residents. Fearing for their lives and property, they knew they were no match for the armed British navy, and many retreated to their homes, barns and ship houses to assess how best to respond. Essex-built vessels, especially privateers, were the targets the British were after. Nearly completed and new brigs, ships, schooners and sloops in the coves, on the ways and moored at the docks were torched. One after another, flames engulfed the hand-hewn wooden vessels, illuminating the night sky and filling the cold air with the acrid smell of smoke and charred wood.

By dawn, the damage was evident. Several vessels still burned and smoldered in the shipyards and coves. In the aftermath, it was determined that all of the twenty-six vessels torched were severely damaged or total losses. The ship *Superior*, built in Essex, was captained by a young man from Lyme, Connecticut, named Henry Champlin and anchored in Middle Cove. Captain Champlin was one of many who awoke in Potapaug to find his ship ablaze. The vessels burning in the coves at dawn were not the end of the mayhem. Some British troops headed to the Falls River, where they knew a large hull was on the ways.

Other marines headed straight for the town. Stores were ransacked, and ship supplies were taken or destroyed. Cordage from the old ropewalk was stored in the basement of the Hayden Starkey Store. The British confiscated a considerable amount of it and demolished the rest. They snatched supplies on their way out and several casks of rum.

The local citizens fought back as best they could. Some tried to douse the burning ships with buckets of water. One group of villagers destroyed several hogsheads of rum so as not to allow the British to take the liquor. Others made their way south on either side of the river with firewood to await further instructions knowing the British had only one way out; they must pass through the narrows.

Captain Coote ordered two of his lieutenants to take as prizes two newly built vessels, *Young Anaconda* and *Eagle*, both privateers. The former was fitted with eighteen guns and the latter with sixteen guns. Both vessels were laden with supplies and hauled downriver by the British. Then, as luck or providence would have it, the winds shifted and *Young Anaconda* went aground. Captain Coote ordered his men to transfer the supplies to the smaller privateer and torch the grounded schooner.

Villagers were outraged and not standing idly by. They joined forces with their militia. Word was sent to New London and to militias in neighboring communities requesting immediate assistance. Saybrook, Lyme and Westbrook all answered the call. The joint militias assembled along both sides of the river and searched for appropriate sites to set up their armaments. British marines were ordered to shelter below the bulwarks or railings on the sides of their boats and to remain low until nightfall. As dusk fell, the British burned the privateer *Eagle* and prepared to make their escape with the outgoing tide under cover of darkness.

Citizens stacked and lit mountainous bonfires to illuminate the British craft as they attempted to slip downstream, but visibility was low. The moon waned to a thin, pale arc and offered no light. To make matters more challenging, a thick mist rose from the river, making British targets nearly invisible to those on shore.

Connecticut citizens gave it their all. Local militias were bolstered by the arrival of troops from New London with several cannons. One six-pound cannon was secured in a strategic location between Ayer's Point and Watrous Point, where the river was the narrowest and their chances were best for reaching the enemy's vessels. Around 8:00 p.m., five hundred people, including townsfolk, militia, sailors and marines from New London, fired toward the river from both sides as the British boats passed by with muffled oars in the darkness. No one could be certain if targets were hit. In the end, the British craft made it silently through the narrows, past the unmanned Saybrook Fort and out to the safety of their warships, but two British marines were killed and several others wounded.

There were no fatal casualties from Potapaug, but the trauma of the raid and loss of twenty-eight schooners, sloops, brigs and ships were devastating. Richard Hayden recovered only $1,300 from *Black Prince*, worth $13,300 before the raid. Brig *Hector*, worth $15,000, was a total loss, as were brig *Felix*, valued at $12,000, and a new schooner and sloop *Washington*, valued together at $9,000.

Judea Pratt's vessel was one of only two vessels spared, purportedly due to "mystic ties of free masonry," according to J.H. Beers's 1884 account. Judea Pratt was a charter member of the Mount Olive Masonic Lodge, Number 52, constituted in 1812. Nathan Pratt, a jeweler, made the jewels for the lodge. During the raid, Captain Coote ordered Pratt's vessel to be burned, but Judea addressed the captain as one Mason to another. The two men conversed for a few minutes, and the captain and his troops did an about-face, leaving Judea's vessel totally intact.

Samuel Williams rounded up some of his shipyard crew. The British had entered the Falls River cove to set fire to the massive *Osage*, but David Williams's yard crew knocked off the dog shores and sent their big, unfinished hull down the spillway and into the river in an effort to save it. Their attempt was to no avail. The hull was set ablaze and burned in the river despite the valiant efforts of young Austin Lay, who swore he repeatedly tried to extinguish the flames. *Osage* was severely damaged. The burning of that prized hull resulted in massive outrage and sorrow and an $8,500 financial loss to David Williams's yard.

Osage's burnt hulk was towed to the North Cove, where it remained partially visible near the gristmill for many years, an eerie reminder at low tide of the dreadful 1814 raid. In 1934, the massive *Osage* keel was exposed and hauled from the cove by a man named Samuel Hunt. Other relics of the 1814 British raid on Essex may be found at the Connecticut River Museum.

Peace talks between the United States and Great Britain began in August 1814, but the conflict did not end until December of that year. It was February 1815 by the time the Treaty of Ghent was signed by both countries, and the War of 1812 officially ended.

The British raid traumatized many Potapaug residents, but for others, life continued even after the substantial affronts and losses. Some people work harder to reach their goals in spite of adversity, their spirits forged a little wider and stronger by the experience. By 1815, many Potapaug residents had moved forward with their lives. George Harrington married Phoebe Lay, Samuel's sister, and the newlyweds moved into their new home on Pratt Street, the name given to New Street. That same year, Captain Henry Champlin, who lost his ship *Superior* in the raid, became a pioneer shipmaster in John Griswold's Black X London Line. Both Henry Champlin and John Griswold hailed from Old Lyme, but Henry had lived in Potapaug for a few years. In 1815, he married Amelia Hayden, granddaughter of Ebenezer Hayden, and her inheritance no doubt helped the young couple build their palatial home in 1818 in what was then the center of town, now known as Champlin Square.

Captain Champlin commanded several ships before 1822, including *Venus*, *Adonis*, *Rubicon* and *Cincinnatus*. Amelia Hayden Champlin tried to enjoy life at sea, but she was plagued with seasickness and often under the weather. For the first decade of her marriage, she accompanied her husband on many

Samuel Hunt sitting on the keel of ship *Osage* after hauling the keel on shore from the North Cove. *Courtesy of the Essex Historical Society.*

Captain Henry and Amelia (Hayden) Champlin's Home. *Paul Foundation, Essex, Connecticut; photo by Jody Dole.*

of his voyages from New York to London and back, and the couple's three surviving children were all birthed at sea.

Some residents of Potapaug, especially those who lost ships in the raid or relied on the shipbuilding industry for their livelihood, never fully recovered financially or emotionally from the brutal attack on the town. Richard Hayden suffered substantial financial losses amounting to $60,000. When added to the trauma and stress of the raid, the losses may have hastened the shipyard owner's death. Richard Hayden died at age forty-four in 1816.

Portrait of Amelia Hayden Champlin (Mrs. Henry L. Champlin). Oil/canvas by Louis Ducis (French, 1775–1847). Painted in Paris, 1818. *Paul Foundation, Essex, Connecticut; photo by Cultural Preservation Technologies.*

Two years later, Richard's uncle Ebenezer Hayden, the wealthiest merchant and financier in Potapaug, died. Since his wife and children preceded him in death, Ebenezer's massive fortune was split between his five grandchildren, one of them being Amelia (Hayden) Champlin. Ebenezer's dream for New City Shipyard was manifested through the efforts of others. As if willed by the spirit of Ebenezer, the shipyard arose like a phoenix from the ashes of the raid, despite initial changes in ownership.

Beginning in the 1790s, a six-decade-long Hayden and Williams New City Shipyard collaboration existed, in one form or another through the 1850s. Malcarne wrote, "The shipyard was owned by Ebenezer Hayden and Samuel Williams," and perhaps after 1803, a formal agreement was written between them. On the 1819 survey map, the New City Shipyard has a notation that reads, "formerly owned by Calvin Williams." Calvin obtained nearly full interest in the shipyard, wharf and store in 1816, but he sold his interest within two years. At that time, Calvin's son was a year old and his wife's Southworth family and her father's shipyard were located in Deep River. These considerations may have had some bearing on his decision to sever ties with New City's yard and invest in the Southworth yard with his wife's Deep River family.

By 1819, Ebenezer Hayden's grandson Charles Uriah Hayden owned one-quarter interest in the New City wharf, and Samuel Williams owned the other three quarters. The shipyard interests, however, were split equally between Henry L. Champlin and Erastus Williams, but Erastus left the shipping industry and Essex by 1825. He moved to Norwich, Connecticut, and established a successful woolen goods manufacturing company. His son E. Winslow Williams inherited the company upon his father's death in 1867 and went on to become a wealthy proprietor and leading citizen of Norwich.

Historian J.H. Beers wrote that the first owners of the second ropewalk, Gurdon Smith and the Post brothers, sold the cordage company to "Hayden, Williams & Co." Apparently, the Haydens and Williamses had a joint interest in both the New City Shipyard and the second ropewalk. In the case of the shipyard especially, it was a relationship that endured for several generations.

Manufacturing and Shipping

The shipbuilding industry increased soon after the war with Great Britain ended. Businesses that began as fledgling cottage industries at the turn of the century grew into viable establishments and thrived as shipyards recovered. The Pratt Smithy was one business that fared better during the embargo and war than most others that provided products for shipyards exclusively. Diversification of products and markets helped the Pratts survive the fluctuations in shipyard purchases.

The Pratt Smithy diversified its product line but kept to one medium—iron. Even without orders for ship hardware, the smithy was occupied with filling orders for ironwork for wagons and carts, gates, hinges and doorways, fences and farm tools, boot scrapers and hooks, fireplace and cooking utensils and fancy cupboard hinges, scissors and whippletree and yoke hardware. Unlike some businesses that moved away from Potapaug Point and the Falls River, the Pratt Smithy remained where it was firmly established in 1720, on West Avenue in Champlin Square.

John Pratt began the Pratt Smithy in 1678 while living in Saybrook. His son John Pratt Jr. learned the trade and was a part-time blacksmith. He also built the Pratt homestead in Essex and moved the forge to his property in 1720. John's son Lieutenant John Pratt was a third-generation blacksmith and is credited for establishing the blacksmith shop in Champlin Square, where he also resided until his death. His son Asa lived there too and worked as a blacksmith in the shop for half a century. Asa's son John and grandson Elias followed in succession as proprietors of the shop. Elias assumed the business in 1827.

Buoyed by the shipbuilding industry's increasing commissions for hardware, the Pratt Smithy, like many shops in town, was once again flourishing. At one point, the Pratt property was a thriving twelve-acre farm and included two barns and the home that now houses the Essex Historical Society. In 1844, the shop needed more space, so Elias Pratt built a new, larger structure from bricks to house his shop, his wares, items to be repaired and equipment.

In all, nine generations of blacksmiths from the Pratt family worked in the blacksmith trade consecutively for 268 years. The Pratt Smithy in Essex remained in operation until World War II, when James Pratt faced a dwindling supply of iron and was forced to close up shop.

The Pratt smiths worked in one place for nine generations, but many others made the decision to leave Essex and grow their businesses in places like Deep River, Hartford and Norwich. Erastus Williams left Potapaug in order to expand his businesses and collaborate with others working in the woolen manufacturing industry in Norwich. Silversmith and inventor Deacon Phineas Pratt invented a device for cutting combs in 1798. Pratt's invention allowed comb makers to produce many more combs per day, increasing both income and inventory. No longer did artisans have to spend all day sawing the teeth of combs by hand from cow and ox horn. Pratt's mechanical device greatly sped up the cutting process, and it could be used to make combs of ivory. Phineas Pratt Jr. and his brother-in-law George Reed had a shop and dam in Deep River and made combs of elephant ivory using Phineas Sr.'s invention.

Ezra Williams's ivory and bone-cutting shop in Meadow Woods was also greatly enhanced by Phineas Pratt's invention, and his business quickly flourished. The young entrepreneur realized he needed more space and more power, and by 1810, he made plans to move his shop to Deep River. With the Falls River powering several businesses at the time, he had no choice but to move to a location where water could provide more power. He moved his ivory and bone-cutting business from the Williams business complex to Deep River in 1817. Deep River was named for the stream that bisects the town, and travelers crossed the river on their way to Hartford and Windsor. Its source is in Winthrop, and it runs downhill into the Connecticut River.

About the same time that Phineas Pratt's invention was mastered by comb craftsmen, trade connections between England, the United States and the Netherlands were established with Eastern Africa and Zanzibar, the main island in the Tanzanian archipelago. Ivory became readily available, business contacts were established and shipping routes mapped out.

Today we shudder to think of the suffering and loss of both enslaved Africans and African elephants due to the demand for ivory products. In the early 1800s, people had different perspectives on the ivory trade and the dangerous manual labor required to harvest and carry heavy ivory tusks great distances from dense tropical inland forests and load them on ships. Many practices acceptable in the early nineteenth century, like those associated with the ivory trade, are deemed abhorrent and illegal today. Ivoryton, now a section of Essex, was once the ivory capital of the United States. Combs, toiletries, letter openers, buttons, piano keys and billiard balls are just some of the many items that were manufactured from the dense white elephant ivory during the nineteenth century and early part of the twentieth century in Deep River and Ivoryton.

In Deep River, Ezra Williams joined forces with George Read, and together they manufactured ivory combs and became a large manufacturing company, Ezra Williams & Co. Ezra's untimely death in 1818 resulted in Mr. Read assuming management of the company, and the name was changed in 1829 to George Read & Co. By 1862, the company merged again and was called Pratt, Read & Co. of Deep River, one of the two largest manufacturers of ivory products made in the United States.

The other notable manufacturer of ivory products was Comstock, Cheney & Company, located in West Centerbrook, now Ivoryton, located about three and a half miles from Deep River. Rufus Greene, a shipping magnate from Rhode Island, employed his son-in-law George Cheney, an ivory trader who managed the shipping operation for Greene in Zanzibar until 1862. At that time, George and his wife returned to Essex with their two young sons, and George Cheney collaborated with ivory manufacturer Samuel Merritt Comstock. His desire to retain and guide his employees led to the creation of a village with housing, a store, a school and recreational areas for his employees and their families. Everything they needed was available in the village.

Calvin Williams left Potapaug Point soon after his brother Ezra. Captain Calvin Williams was master of the sloop *Providence*, brig *Pocahontas* and schooner *Transit* in 1815 and schooner *Centinel* in 1819. Calvin married Nathan Southworth's daughter Eunice, and the couple's marriage cemented ties between the Williams and Southworth shipbuilding families and allowed Calvin to become a very successful shipping merchant.

The prominent Southworth family was well established in Deep River and owned three businesses: a sawmill, a lucrative granite quarry and the shipyard. Brothers Job and Nathan Southworth were building ships at

Granite home of Captain Calvin and Eunice (Southworth) Williams. *Courtesy of the Deep River Historical Society.*

their Southworth yard all through the 1790s, including the sloop *Hannah* and 119-ton brig *Rowena*. Calvin's aim was to develop his father-in-law's Southworth Shipyard and repair business. Like his grandfather, father and brothers, Calvin was industrious and knew a lot about shipbuilding, sailing and shipping from his family's many Meadow Woods enterprises. Deep River shipyards produced about sixty wooden vessels before the shipbuilding industry started to decline.

In the mid-1820s, Calvin Williams had a substantial granite home built at the base of Kirkland and River Streets near the Deep River steamboat landing. The Southworths' quarry business produced magnificent granite in shades of yellow and gray as may be seen in Calvin and Eunice's lovely home today.

Southworth granite was probably shipped from Pratt's Wharf by way of water to the town landing. Only two houses in Deep River are made from this granite, as most of it was shipped to other communities. Deacon Ezra Southworth built the other granite edifice in 1840 as a wedding gift for his bride, also named Eunice. Their stately home was left to the Deep River Historical Society by their granddaughter Ada Southworth Munsun.

Fellow Deep River shipmasters built their houses near the waterfront also, including Samuel Mather and Joseph and Justice Arnold. In time, a neighborhood of ship captains and their families developed near the Deep

Denison homestead and shipyard, Deep River, Connecticut. *Courtesy of the Deep River Historical Society.*

River waterfront and landing, all around Kirtland Street, River Street and Phelp's Lane.

When Job and Nathan Southworth retired from shipbuilding, Thomas and Eli Denison, a father-and-son team of shipbuilders, took over the work of designing, building and repairing ships at the Southworth yard.

Calvin Williams promoted the shipyard, and the business flourished due to hefty commissions for vessels from wealthy entrepreneurs. These men included ivory merchants Alpheus Starkey and Joseph and Samuel Comstock and Alvin Whittemore, who manufactured a product called witch hazel, a lotion created to heal and soothe skin. Whittemore was another successful entrepreneur who began his operation in Essex and moved to Deep River in order to expand his business. Like Calvin Williams, Alvin Whittemore remained a resident of Deep River and grew his businesses there but maintained ties with Essex family and friends.

Captain Williams built one of the first marine railways in the state and purchased 110 acres with two houses included from his father-in-law, Nathan, and brother-in-law Nathan, Jr. Ultimately, Calvin Williams's property included both the Deep River waterfront of today and the entire Mount Saint John School property.

Ties between Deep River and Potapaug Point remained strong due to family connections in both locations. During the first decades of the nineteenth century, shared professions and strong blood ties cemented relationships between villages. To many folks, including the Williamses and Posts, whose family members resided in both villages, Deep River was considered a section of Essex.

Canals, Shipping and Steamers

In May 1820, the Borough of Essex was created by an act of the Connecticut state legislature. Potapaug Point then became known as the Essex Borough of Saybrook, and it was not until 1854 that Essex Borough and Old Saybrook were separated and the Town of Essex was established. Three villages or districts are included in the town: Ivoryton, Centerbrook and Essex.

Samuel and David Williams's shipyard in Meadow Woods thrived in 1821. Calvin Williams commissioned his family's shipyard on the Falls River to build a new vessel for his use. Calvin's brother-in-law and master carpenter Captain Richard Hill and several of the Williams brothers accepted the offer and built the new ship for Calvin. All of the Williams brothers were shipmasters except for the youngest, Richard Pratt Williams, who at age twenty-five in 1821 was well on his way to becoming a master shipwright. The name given to the new vessel was *Six Brothers*, and the Williams brothers operated the vessel out of New York for mercantile purposes until 1827. Captain Calvin Williams had command of *Six Brothers* during that entire six-year period.

The year 1822 was a difficult one for the Williams family. Business boomed in all of the Williamses' shops surrounding the Falls River, but the patriarch of the family and founder of most of the enterprises, Samuel Williams, died that year at the age of age seventy-one. Samuel's sons were all well into their respective careers by then, but all worked with their father much their lives and surely felt the full weight of his passing. The shipyard was left to Samuel's son David, who had taken over the yard's operation. David soon built a new mill at the site of his father's old sawmill and continued his shipyard's output of about two vessels per year, mostly under three hundred tons.

Samuel Williams's interest in the New City Shipyard was left to his son Calvin, who sold the interest to his brother Richard P. in 1826. Calvin was established in Deep River and had a new house for his wife and son Richard C. New City Yard was no longer co-owned by Erastus Williams, as he moved to Norwich that year, but Henry Champlin was an owner of the yard, and Richard Pratt Williams was the yard's primary shipwright for the following two decades.

Attention was drawn to New York in the early 1820s, as news of construction on the Erie Canal made its way to Essex Borough and beyond. Governor DeWitt Clinton of New York was credited with initiating the inventive canal project, which met with a fair amount of opposition. Clinton dispelled the naysayers who disliked the plan or those who thought it impossible to accomplish and called it derogatory names. The governor persevered with his ambitious plans. Beginning in 1817, the New York State Commission oversaw the development of a canal system that allowed travel and trade vessels to navigate from Lake Erie to the Hudson River. This canal system opened the country's midwestern territory to encourage settlement and commerce. The four-foot-deep, forty-foot-wide canal was dug from Albany, New York, to Buffalo, New York, with a series of cement locks strategically placed along the 363-mile system. The locks were designed to lower or raise vessels as needed while traveling to or from the Hudson River inland or on return.

Several New York companies dug sections of the ditch, and thousands of men labored to construct the great Erie Canal. Over one thousand men lost their lives during the seven years of canal construction, most from illnesses related to working in swamps, collapses, accidents and explosions while using gunpowder. The loss of these laborers was a high price to pay for turning New York into a significant financial center and opening the upper Midwest to development. On opening day, though, most folks looked to the future. With bands playing and cannons lining the riverbanks, the Erie Canal opened on October 26, 1825, with great fanfare and a flotilla of ships. New York governor DeWitt Clinton led the parade of vessels down the Erie Canal with a cannon salute. Guns fired consecutively from both sides of the river all the way down the canal and back up to Buffalo, New York, later in the day. Folks from Connecticut and other states went to New York to celebrate the great engineering feat that made New York a major center for commerce.

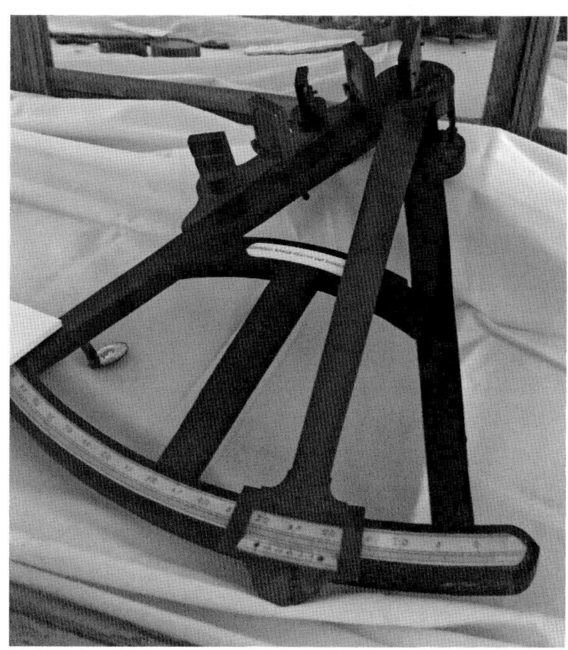

Joseph Post's quadrant, taken to sea from 1812 to 1836. *Courtesy of the Deep River Historical Society.*

One man from Essex who took part in the opening-day flotilla was seasoned Captain Joseph Post, son of Captain David Post and Lucretia Post, descendants of Stephen and Elinor Post of Old Saybrook. Joseph Post went to sea as a boy with Samuel Williams's son Captain William Williams on *Laguira* out of New York. By 1814, Joseph Post had worked his way up the ratlines and took command of *Laguira* at the age of twenty-five. Subsequently, he captained several ships over the course of the next twenty years. His early commands included the brig *Pocahontas*, in 1815; ship *Mirror*, in 1818; brig *General Scott*, in 1820; and new ship *Henry Hill*, built in Essex in 1822. New ship *Extio*, 278 tons, was built in Essex in 1825, probably at David Williams's yard since Joseph had close ties with William Williams. *Extio* is most likely the ship Captain Post sailed in the October opening day flotilla in celebration of New York's Erie Canal.

Captain Joseph Post's maternal grandmother was Mary Denison, and her line went back to Captain George Denison and his wife, Lady Ann Borodell of Stonington, Connecticut. Joseph's father, Captain David Post, was born in 1764 to Deacon David Post and Deborah Ward. Captain Post sailed Essex-built ships in the coasting trade for the Chesapeake Line. This line of ships ran between New York and several ports in Virginia and Washington, D.C. It also was the first line to establish a steamship route between New York and the Chesapeake Bay area. In 1822, Captain David Post was in command of the schooner *Mark Time* and earned his livelihood through merchant trading along the East Coast, a profession he passed on to his sons Joseph D. and David Rawson.

Captain David and his wife, Lucretia Post, had a family of eleven children that began in 1789 with Joseph D. Post's birth. Four children died in infancy. Four others died in their twenties, but three of their children survived to adulthood: Joseph, born on March 21, 1789; Olive, born on January 16, 1791; and David Rawson, born on August 20, 1807, eighteen years after his brother Joseph's birth. If it were not for their sister, Olive, who skillfully drew the Family Register of David and Lucretia Post, we might never have known of Olive or her eight siblings who died young.

Olive's illustrated family register of her immediate family was drawn in a Dutch Fraktur design, using traditional stylized motifs of birds, tulips, angels and hearts, but her Fraktur also includes four female figures, one in each corner, representing the values of their family: peace, hope, charity and love. Olive rendered her design in pen and colored inks, including a white pigment that is still bright today, nearly two hundred years later.

As it was then and is now, women of the family are the keepers of family history. When wars, floods, fires or children with scissors destroyed birth, death or marriage records and other important documents, women would make certain that copies of the vital statistics were accurate and passed down in the family as heirlooms. Family records took many forms. Some were hand-drawn and colored like Olive's or written in perfect, tiny script on fan charts of seven to ten or more generations. Others were drawn as family trees or embroidered on linen samplers, and from the seventeenth to the nineteenth century, many family records were recorded in center pages of large, leather-bound family Bibles, and these treasured books were often taken to sea.

The fact that Olive Post's family register and her parent's family Bible survived to this day in any form is somewhat miraculous. Olive married Captain John R. Rockwell, son of Jabez Rockwell, born in 1762, and Irene Porter of Tolland, Connecticut. Unfortunately, Captain John R. Rockwell was lost at sea on September 13, 1824, at age thirty-three, but not before having a daughter, Mariah, who died at three months old, and two sons, John E. in 1815 and James P. in 1817. Captain Rockwell descended from Sir John Rockwell, who, according to family lore, rescued the Earl of Northumberland and Lord Percy from the party of Earl Douglass at the Battle of Halidon Hill in 1333. Sir John Rockwell was knighted for his successful rescue mission.

Five generations of Post family Bibles. *Courtesy of R. Major.*

Following her husband's death, Olive Post Rockwell took her young sons, John and James, to her mother's Post family homestead in Deep River by Devil's Wharf. Olive brought the boys up by the river with Post Cove directly behind the homestead. Her treasured family register was subsequently damaged over the years and disintegrated to pieces, with one section snipped by a child as if to make a snowflake. All that was salvageable was tucked into a small envelope for five generations before it was finally pieced together again in the 1960s by Olive's fifth great-granddaughter for her grandmother Marjorie Post. After moving the pieces around several times in an effort to make sense of it, she finally read the adage boldly written across the top of the old mid-1820s Fraktur signed by Olive Post Rockwell. The heading, written in bold letters, was, "Keep Sacred the Memory of Your Ancestors." The Post family register reminds us today how important it is to document family history and to honor the lives, strengths and accomplishments of our ancestors.

As early as 1822, Nathaniel L. and George Griswold developed a shipping line from New Orleans to the West Indies and, later, to China for tea. They added a California run for those seeking a fortune in the gold rush and commissioned the legendary clipper *Challenge*. Its epic voyage from New York to San Francisco is one of the greatest sagas of the clipper era. In 1823, John Griswold of Old Lyme began a shipping line that would become his Black X Line of London packets out of New York, initially operating in conjunction with Fish, Grinnell & Company. In 1827, Griswold parted ways with his partners and operated his own company with four ships. This line became extremely popular, and more ships were added. By 1852, the Black X Line had eight ships and eight regular shipmasters, most from Essex and Lyme, making regularly scheduled trips from New York to London. Fish, Grinnell & Company took its ships and started the Swallow Tail Line of packet ships to Liverpool beginning in 1827, competing with the Black Ball line for Liverpool service.

One of John Griswold's pioneer masters was Captain Henry Champlin, who commanded the new ship *Hudson* for the line in 1824. He also commanded *Hudson* in the Swallowtail line in 1827. It appears that Captain Champlin returned to the Black X Line and ran a series of packets for that line, including ships *Cambria*, *Sovereign*, *President*, *Philadelphia*, *Montreal*, *West Minster* and his last command, *Mediator*, in 1836. From then until 1840, Captain Champlin was superintendent of Griswold's Black X Line.

Another early shipping company, E.D. Hurlbut & Company of New York, had its sights on the growing cotton trade. In 1825, it set up an office on

Post family register, by Olive (Post) Rockwell, 1820. *Courtesy of R. Major.*

South Street in New York with agents in destination cities including Mobile, Alabama, and later Pensacola and Apalachicola, Florida.

About the same time as the coastal packet lines of John Griswold and E.D. Hurlbut and others were getting underway in the mid-1820s, the establishment of early steamship lines began rapidly transforming the manner in which people traveled over the water. Deep River's steamship dock accepted steamship dockings in the 1820s as did Essex, and by the 1830s, steamship lines ran up and down the Connecticut River making several stops along the way. Trips to New York, which could take up to five days by sail due to weather conditions, took two days by steamship. People were able to travel to New York from Hartford in a fraction of the time it took to sail there.

Serious accidents occurred on the early steamships due to fires, collisions and boilers bursting and scalding and maiming people, but in 1852, the Steamboat Act was passed, which greatly improved safety aboard steamships. Inspections, construction standards and plans for reducing fires and collisions were required of all steamships.

E.D. Hurlbut & Company developed a line of coastwise packet ships that could provide regular service to southern ports, where heavy bales of raw cotton were loaded aboard ships by enslaved dock laborers called stevedores, forced to stow heavy bales of cotton into the dank hulls of ships.

Well acquainted with the East River and Port of New York, Captain Joseph Post joined forces in 1825 with E.D. Hurlbut & Company to pioneer its fledgling coastal packet line to Mobile, Alabama. Mobile was the second-largest port next to New Orleans for cotton. Soon, Essex packet masters, related by blood and or friendship, were employed in the new shipping company. The world was opening up, not just for shipmasters but for residents as well. Steamships brought folks to new towns, and residents were drawn to new parts of the country. Olive Post moved on with her life. Her son John Rockwell went off to sea as a ship's boy, starting out his career most likely under his uncle Joseph Post's tutelage. Olive met Lemuel Butler from Rochester, New York, and they were married on June 13, 1830. The couple moved to Cincinnati, where Olive Post Butler died a decade later.

Shipmasters employed by E.D. Hurlbut's Mobile Line in the late 1820s and early 1830s included Joseph Post, Alva Post, Cornelius R. Doane, David Rawson Post, brothers George and Frank West, John Urquhart, William H. Williams, Ezra Denison Post and Russell Handy Post, cousins of Joseph and David Post. Early ships in the line included *Extio*, *Edwina*, *Lorena*, *Tuskina*, *Albamian*, *Russell Baldwin*, *Elisha Denison*, *St. John* and *Hector*, built in Essex for Captain Joseph Post in 1833. In fact, several of these ships were built in Essex. New ships were commissioned by E.D. Hurlbut from New York and Essex shipyards, initially for use in the coastwise cotton trade and later for transatlantic lines. At first, the baled cotton was brought up to northern mills and factories for processing, but by the early 1830s, E.D. Hurlbut &

Company also became a transatlantic business, working a triangular cotton trade involving southern ports along the lower East Coast of the United States, New York and Liverpool. Imported goods, finished cotton and passengers were carried on homeward-bound trips. In 1830, an estimated fifteen thousand immigrants sailed to New York from Liverpool. Great Britain purchased over half of the American cotton exports and opened several factories in Lancashire to spin and weave the cotton into cloth. The Hurlbut line became more regular and predictable, and the coastal cotton trade grew. By the 1850s, there were several more European lines, with stops in Liverpool, Le Havre, Antwerp and Rotterdam. Returning ships sailed to New York with European goods in the hold, immigrants in steerage and cabin passengers on the top deck with saloons and individual first- and second-class cabins.

Captain Francis West, known as "Frank," moved to Essex in 1825 from New London. His objective was to pick up his new schooner, *Jesse*, built for him at David Williams's yard by master builder Richard Hill. Captain West remained in Essex and built a home near Williams Shipyard. He took command of several ships built in Essex. These included new ship *Elisha Denison*, built in 1827; new ship *Russell Baldwin*, in 1833; and new ship *Edwina* in 1837. These ships were all packets built in Essex for use in the E.D. Hurlbut & Company's Mobile cotton trade.

Early packet ships like *Elisha Denison*, built in the 1820s and displacing 359 tons, may have been one of the transitional packets that were collaborative ventures of Williams Shipyard and New City Shipyard, but ships over that tonnage were probably built entirely at New City. If the ships were too long or too heavy and their draft too deep in the water, mariners would not have the proper depth and width in the Falls River necessary to float or maneuver the hulls.

David Williams's brother Richard P. Williams most likely took over the shipbuilding at New City soon after he purchased the Williams interest in the yard from his brother Calvin in 1826. Richard was thirty years old by then, married to Louisa Bushnell for ten years; by all accounts, New City began to thrive under Richard's direction. Captain Henry Champlin was an owner of the yard at that time, but whether there was a co-owner after Erastus Williams moved to Norwich is unknown.

Captain Champlin may have been the sole owner of the shipyard in 1826, though not owner of the New City wharf and store. As New City yard owner, Captain Champlin would have managed the accounts of the shipyard operation, while Richard Williams, master shipwright, attended to

Captain Henry L. Champlin. Watercolor by Prescott & Gage (Hartford, Connecticut, active 1861–65). *Paul Foundation, Essex, Connecticut; photo by Cultural Preservation Technologies.*

the actual design and building of ships and overseeing the day-to-day work of the ship specialists, including ship joiners, caulkers and ship carpenters.

With the availability of ship's hardware from both Judea Pratt's forge near the end of New City wharf and his Williams forge on the Falls River, Richard Williams had easy access to iron fittings, nails, chains, hooks, hinges, cleats, oarlocks and other iron hardware. Lumber from the Williams sawmill and from other mills in town provided the boards and timbers required for constructing each vessel. In the case of lumber, it could be floated down into the cove to New City yard from the Williamses' Falls River lumberyard. If more pilings were needed after a storm to shore up the wharf, logs soaking on the south side of the Falls River could be taken from behind the water fence and floated down into the cove. By 1830, both the Williams yard under David Williams and the New City yard under Richard Williams thrived, with ships on the ways and others being finished off at the docks and landings.

Another change was taking place at a shipyard off Mack Lane, formerly the Hayden Shipyard. Uriah Hayden sold his yard to his nephew Calvin Hayden, who built ships there until 1830. At that point, he sold the yard to a young shipwright named David Mack, and the name of the yard was changed to Mack Shipyard. David's family members and employees who worked in his yard lived in the houses surrounding the yard and Middle Cove. David Mack's homestead was constructed at the base of Mack Lane beside his yard. He built coasting schooners, including *Zerviah* in 1839, and later added a second yard on the west side of Middle Cove. A prolific master shipwright, David Mack built schooners for nearly four decades into the early 1870s.

In 1829, Calvin Williams was captain of the schooner *Texas*, built by his brother David in the Williams yard in Essex. This schooner was used to pioneer the Texas trade. In 1830, Thomas and Eli Denison took over the operation at the former Southworth yard. The next year, Captain Williams was master of brig *Jane* and sailed it in the new Texas line also. That was his last command, for Calvin Williams left the sea and retired to his home in Deep River. He gave his captaincy to his nephew Captain Frederick Williams, and within a year, Calvin Williams died at the age of forty-seven on August 21, 1833. Along with other items in Calvin's will was his marine railway near the shipyard that carried on business until the 1870s by the Denisons, cousins of the Essex and Deep River Post family.

As one shipmaster ended his career at sea, another ship's boy started his long journey to the quarterdeck. John Urquhart entered his profession like many poor lads in Liverpool and New York. He was barely a teen when orphaned at age thirteen and soon after went to sea with Captain Joseph Post. Urquhart came up through the hawse pipe like countless other boys, and by 1825, at age twenty, he was chief officer under Captain Post on schooner *Henry Hill*. Joseph Post's cousins Captains Alva Post and Ezra Denison Post also commanded *Henry Hill*.

When young John Urquhart met Captain Joseph Post in New York, he may have been orphaned, but he was not uneducated. Urquhart was the son of an immigrant father by the same name who grew up in Findon on the Black Isle, on the banks of Cromarty Firth in the Highlands of Scotland. He graduated from Kings and Marischal Colleges in Aberdeen with a master's degree in the classics and immigrated to the United States in the late 1780s. In 1791, he accepted a position as headmaster at New-Ark Academy. Two years later, he was ordained as an Episcopal rector in New York City and was sent up to the Highlands of Amsterdam and Johnstown, New York, to minister to Episcopal churches there. Reverend John Urquhart married Susanna Van Vorst, A Dutch-Irish girl of Schenectady and Johnstown, New York, where her grandfather Robert Adems brought her up after both parents died. Adems was Sir

William Johnson's accountant, and he handled all of the baronet's trading enterprises and store until his death in 1774.

Both Episcopal parishes in Amsterdam and Johnstown were decimated during the American Revolution, and life was harsh for Tory sympathizers for years after the war ended in 1883. Reverend Urquhart struggled to make ends meet as the Urquhart family grew, and he was forced to open a school for boys in their parish home to make ends meet. The boarding school included meals and laundry, adding significantly to Susanna's daily chores.

In 1818, Reverend Urquhart's eldest son, John, faced an uncertain future. Living with four older and younger siblings and other tenants in a hot, cramped apartment in Brooklyn was not nearly as alluring as Captain Joseph Post's schooner, moored at E.D. Hurlbut's dock off South Street in New York. Joseph Post took several lads to sea over the years and trained many of them to be shipmasters. Young John received his formal education at his father's private home school for boys, but his practical knowledge of celestial navigation, sailing and essential skills like the proper use of nautical instruments, charting a course and effects of tides and winds in different locations and seasons was learned from being at sea with Captain Post. A decade after his mother died, John Urquhart became a shipmaster in 1828. He was given command of *Henry Hill* and welcomed into the clan of shipmasters, mostly from Essex, who were employed by E.D. Hurlbut and Company. Joseph Post's brother David R. Post was two years younger than John and striving to become a master in Hurlbut's packet line also.

At some point around that time, David Post was introduced to John's Urquhart's younger sister Meriah. She was just five years old when their mother died, and the youngest in the Urquhart family, born in 1813. Her much older sisters Isabella and Catherine took care of her and her brother Walter in Brooklyn, but when Meriah was about eighteen years old, her brother John moved to Essex and brought Meriah with him. John and Meriah were very close siblings and decided to establish their lives in Essex. The relationship between Joseph Post and John Urquhart was a fortuitous one with far-reaching consequences, as will become evident.

Joseph Post and John Urquhart continued to be regular masters in E.D. Hurlbut's line of coastal packet ships, sailing to southern ports to pick up cotton and then bringing the bales up to northern mills for processing. In the early 1830s, Joseph Post pioneered E.D. Hurlbut's transatlantic line to Liverpool and developed contacts there. He was definitely familiar with the port by 1834. Joseph was referred to in family discussions of ancestors as "an old Liverpool Captain with a rather scurrilous reputation."

Liverpool was part of the Lancashire County in the mid-nineteenth century. According to an excellent online exhibit from the Merseyside Museum in Liverpool, England, titled "100% Cotton," the nineteenth century cotton trade was a lucrative business. Raw cotton was brought directly to Liverpool, where it would be sold to cotton merchants who assessed the bales and made deals right on the dock during the first years of the trade. As this business grew, companies were forced to open several cotton mills in Lancashire to process the raw cotton, mainly from other countries, including the United States and India. Working conditions in the mills were often hazardous for the workers, including children, who labored long hours. Transatlantic trade became as common as coastal trade and was often an expansion of the coastal trade. At one point, E.D. Hurlbut & Company had thirteen lines, including four to European ports.

While many shipmasters changed shipping lines, it appears that several masters who worked for E.D. Hurlbut remained with the company throughout their maritime careers. This was true for Joseph Post, David R. Post, William Williams, Cornelius R. Doane, John Urquhart and Ezra Denison Post, among others. Driving packet ships in all seasons took true grit and intelligence, and these men rose to the challenge. The same held true for the shipbuilders who took years to master their craft and built their ships out in the elements and exposed to all types of weather. Careers could last from ten to thirty years in either profession. The exemplary qualities demanded of shipbuilders, yard owners and shipmasters made these men highly esteemed members of their community.

During the nineteenth century, several prominent men in the lower Connecticut River area made their fortunes as merchants, builders, shipmasters and shipping agents. These same men, when not at sea or required at their shipyards, took on civic projects in their community with the same determination, commitment and passion. Take for example, the Essex borough needing to fund and build a new school in 1831. This project required financing, construction, purchasing land, hiring a headmaster, ongoing financial support and related tasks. Note the list of charter names of "enterprising citizens" who supported the establishment of Hill's Academy in Essex, which attracted boys from several villages. Prominent citizens included David Williams, Richard P. Williams, Joseph Hayden, Ezra Mather, Uriah Hayden, Timothy Starkey, Gideon Parker, Henry Champlin, William Williams, John Urquhart, Alva Post, Reuben Post, Joseph Post, Noah Starkey and Austin Starkey. Joseph Hill donated the land for the new school, and it was built in 1832.

In the early 1830s, John Urquhart and his wife, Ann Carr, had two young children. The Urquharts' home was located on the corner of Denison Road in the Meadow Woods section, near the Williams complex. Formerly Calvin Williams home, the house was built in 1806 by shipbuilder and shipmaster Tabor Tooker. By September 1833, Captain Urquhart had established his roots in Essex, commanded new ship *Lorena* and was gainfully employed by E.D. Hurlbut & Company as a regular master in its coastal lines.

Alvin Whittemore was refining his process for making a product for soothing skin called witch hazel and married Nathan Southworth's daughter Theresa on October 2, 1833. Like her sister Eunice S. Williams, Calvin's widow, Theresa was from Deep River and the widow of Captain Thomas Masson, a native of Banff, Scotland, who drowned at sea in 1830 off Cape Henlopen, Delaware. Theresa had two children with Captain Masson, eight-year-old Thomas Jr. and June, age eleven. June died two years later, but Alvin and Theresa Whittemore added three more children to the family, and Alvin continued to build his witch hazel enterprise in Deep River.

Six days after the Whittemores' marriage, a tragic accident occurred on the New York–Hartford steamer *New England* just off Essex. The steamer had comfortable sleeping quarters, and many passengers were asleep. Captain Waterman and some of the steamer's crew were engaged in landing a rowboat with passengers from Essex. About three o'clock in the morning, there was a tremendous blast. Both of *New England*'s copper boilers exploded, tossing large pieces of hot iron into the river. Several women were scalded when struck by a blast of steam as they fled their upper-deck cabins. While the citizens of Essex did their best to care for the survivors and a surgeon was sent for, thirteen people died, and several others were badly injured and scalded. In addition to extensive damage to the steamer, passengers' luggage, merchandise and valuables were lost to the river. Boiler explosions were responsible for several types of injuries, including severe bruises, sprains and broken bones from falls or being tossed into walls and hard surfaces, severe scalding and disfiguration from high temperatures of the steam, drowning when tossed into frigid water, burns from fires aboard vessels and emotional scars from the trauma of the experience.

Despite this October 8 catastrophe on the steamer *New England* and similar boiler explosions, steamers continued to be a standard means of travel on the river to and from New York. Eighteen steamboats plied the river between Hartford and New York from 1834 to 1851. The early 1830s was a period of rapid expansion of travel routes on land and sea, of shipping lines and scheduling, of experimentation with newly designed mechanical devices and systems for steam engines. As with all new inventions and mercantile enterprises, growing pains were inevitable, and modifications, improvements and new laws and regulations were necessary, especially with regard to safety issues.

Cotton Trade, New Regulations and Bankruptcy

Captain Joseph Post took command of cotton packet *Tuskina* in 1833 and headed to Mobile, Alabama, on business and remained in port awaiting the arrival of his new ship, *Hector*. His mentor, Captain William Williams, then fifty years old, sailed the new ship from New York down to Mobile to deliver *Hector* to his fellow line captain. No doubt Captain Williams shared his thoughts on how the new ship handled, and hopefully the two old friends had time to reminisce about their early days in Meadow Woods, escapades on land and challenges at sea. Joseph Post and William Williams parted ways without knowing that would be the last time they saw each other.

Captain Williams took over command of *Tuskina* and headed out to deliver the cotton. At some point, months later in the voyage, the captain became ill and died aboard *Tuskina* on October 1, 1835. The ship's crew responded as they had to, considering there was no refrigeration in those days. They brought Captain Williams's remains back home to Essex in a cask of rum. William Williams was buried a month later at Riverview Cemetery.

Joseph Post enjoyed being in command of his new ship, *Hector*, built for him in Essex and commissioned by E.D. Hurlbut & Company. The ship was 557 tons and nearly 134 feet in length. Joseph's brother David Rawson Post worked his way up to being chief mate on *Hector* under his brother's tutelage. The Post brothers' 1834 voyage to Liverpool was destined to be memorable. Having heard a rather dubious family story from Marjorie Post concerning her great-uncle, Joseph Post, Ruth enlisted her daughter

Ship *Hector*. Watercolor by Frederic Roux (French, 1805–1874). Painted at Le Havre, May 1841, for Captain David R. Post. *Paul Foundation, Essex, Connecticut; photo by Cultural Preservation Technologies.*

Paris Major's help to determine whether there was any truth to the tale, as they share a love of research.

As the story was told to Ruth, Joseph Post sailed to Liverpool with a load of cotton, his ship was unloaded and he departed for New York. He was homeward bound when a passing packet signaled for him to heave to. Captain Post stopped his ship and prepared to receive a message from the other ship. A letter passed from ship to ship, informing Captain Post that his wife, Maria, had died in Essex. She was the mother of the three of their seven children who survived, ages eleven, nine and five. As the story continues, Captain Post turned the ship about and headed straight back to Liverpool, where he wed a young woman and brought her back to Essex.

Most family tales are never passed on. Others make their way down generations in jokes or whispers in intimate conversations, simplified or embellished along the way. This family story concerning Captain Joseph Post descended five generations before Paris and Ruth researched it and found it to be nearly completely accurate. Maria W. (Heath) Post from Newport,

Rhode Island, died in Essex at the age of thirty-eight on September 18, 1834. She was buried in Riverview Cemetery, in Essex. Her seventh child, a daughter, died that year soon after her birth, and Post family records written by Robert Denison reveal that while his family lived in Essex, Captain Post was then living in New York. The caption on Maria's headstone reads in part, "Wife of Joseph Post."

Captain Post did not go directly home from Liverpool as planned, and it is not known at this time what became of his three surviving children with Maria. Upon receiving the news of Maria's death, Joseph ordered the ship to return to Liverpool. He located his intended, and with three of her family members present, Joseph Post, "widower from New York," married Mary Prichard in Liverpool, Lancashire, England, on February 10, 1835. Joseph was then forty-six years old and his bride twenty-five years of age. The couple sailed to New York, no doubt with his newly promoted brother Captain David R. Post at the helm much of the time.

Joseph and Mary Post lived in Essex for five years in a house that was built for Charles Uriah Hayden, a grandson and one of the beneficiaries of Ebenezer Hayden. Charles commissioned the grand home on West Avenue near the home of his cousin Amelia Hayden Champlin, but he ran into financial difficulties and eventually was forced to sell the house. Joseph Post purchased the house from Hayden's financial executor and added an ornate cupola, surrounded by several arched windows to the roof of the house.

Mary Prichard Post was very comfortable in her beautiful West Avenue home, as it had several unique features, including a built-in cistern fed by a nearby spring. This feature was no doubt most cherished, as Mary gave birth to a son the next spring, and like all infants, he required frequent bathing and changing. The Posts named their new son Prichard, Mary's maiden name.

Joseph Post and John Urquhart must have approved of the relationship that developed between their younger siblings, then twenty-nine-year-old David Post and twenty-three-year-old Meriah Urquhart. The attractive couple married on August 18, 1836, and their lavish summer wedding in East Haddam united the Post and Urqhuart families for generations to come. The newlyweds had a second ceremony in New York with Meriah's Urquhart siblings, including her brother Captain Walter Urquhart and his wife, Elizabeth; sisters Catherine and Isabella and their husbands, who were brothers; and Captain John, his wife, Ann, and their children. After the celebrations were over, David Rawson Post continued as master of *Hector* and remained a regular captain in E D. Hurlbut's Shipping Line for his entire career.

In 1837, Joseph and Mary Post welcomed another son, whom they named David Rowland Post. Both of their young sons would one day follow in their father's footsteps and become sea captains. David and Meriah Post also had a son in 1837 and named him John Urquhart Post. The couple was devastated when their firstborn died at only six months old in November 1837, but in August of 1838, Meriah gave birth to a daughter and they named her Lucretia Maria Post.

While the number of shipping lines was increasing and shipping company headquarters lined South Street, there were unresolved issues developing between pilots who hailed from New York and those from New Jersey. Following a serious mishap in 1837, an inquiry was held in New York to determine the facts surrounding the incident. Those in charge of the inquiry hoped to gather information from reliable sources to justify pilotage regulations. This would affect all ships entering or leaving the Port of New York.

It was customary for ships to wait in the Lower Bay at Sandy Hook and raise a lamp from one of the ship's yards signaling the need for a pilot, a local person with knowledge of the navigational changes in the local bays, channels and harbors who could escort ships safely through the narrows to the Upper Bay. Pilots were especially needed as ships became larger and the chances of getting stranded on the bar increased. However, prior to new regulations being put into place, pilot boats were not always available for various reasons, resulting in several instances where ships waited hours, even days, for a pilot. The situation was dire during inclement weather or when a ship was in distress and no pilot came to its aid.

Several respected ship owners and masters, including Captain Henry Champlin, provided testimony for the need for changes and improvements in the pilotage system. One incident recorded involved the disabled barque *Mexico* that waited in the Lower Bay with more than thirty square-rigged vessels for a pilot. Despite its distress signal and lanterns hanging from the spars of every ship waiting, no pilot arrived that night. A powerful nor'easter developed soon after midnight of the second day of waiting. Unable to hold to windward, the distressed barque was blown forty to fifty miles in the gale-force winds. *Mexico* went ashore twenty-six miles east of Sandy Hook off Hempstead, New York. Due to the frigid temperature and constant battering of the ship by monstrous waves, 108 men, women and children froze to death clinging to the rigging on the ship. Only 4 men, including the captain, were saved by a local resident and his sons, who risked their lives to save the few from the barque's icy bowsprit so that they could bear witness to the tragic incident.

A visitor who surveyed the scene in an open tavern barn where frozen bodies of the dead were brought for identification after washing up on shore, wrote,

> *There were scattered about among the dead four or five beautiful little girls, from six to sixteen years of age, their cheeks and lips as red as roses, with their calm blue eyes open, looking at you in the face, as if they would speak....I touched their cheeks, and they were solid as a rock and not the least indentation could be made.*

Following numerous testimonies, during and after the inquiry on pilotage complaints, new laws were enacted and changes instituted for both New Jersey and New York pilots. Competition between pilots from both states was permitted, and from then on, the system of pilotage greatly improved. Today, in the United States, there are state and federal laws requiring ocean and coastal vessels to engage pilots for navigation into and out of ports, and licensure is required of all pilots.

By 1838 and for the next several years, Richard P. Williams was at the peak of his career at New City. He built one packet ship after another and launched them into the North Cove, where most of the ships, once fitted out at the end of the dock, headed off for New York to receive cargo and passengers. In 1838, Richard Williams built the packet ship *Floridian* for E.D. Hurlbut & Company. The following year, he enlisted the help of shipwright Noah Starkey, and together, they built the beautiful packet *Elizabeth Denison* for Captain Russell Handy Post. The ship was 644 tons and 142 feet in length. It had a fully carved and painted figurehead of a woman in a flowing gown prominently displayed on the bow of the ship. *Elizabeth Denison* was contracted in 1839 for Hurlbut's new line to Antwerp, and according to historian Thomas A. Stevens, the ship "was the largest ship built there to date." New City Shipyard was thriving due to its competent owner, Captain Henry Champlin, and shipwright Richard P. Williams's excellent workmanship and oversight.

At the same time, conditions in the Falls River were rapidly deteriorating. Richard's older brother David Williams was facing serious financial difficulties. David did what he could to keep his shipyard and the sawmill in operation, but extenuating circumstances beyond his control contributed to mounting business losses. By 1839, the Williams yard had produced about sixty vessels, mostly schooners and sloops less than 350 tons. It had sold many prefabricated vessels and house packages, but business started to decline as the demand for steamers and larger wooden vessels increased. With the loss of trade in the West Indies, followed by the British embargos, financial losses during the War of 1812 and the introduction of steam-powered vessels in the 1820s, the number of commissions for smaller river-built vessels decreased as the century progressed. Shipmasters and shipping lines ordered larger ships from New City's yard, Redfield and Parmelee's yard located directly south along the cove, and they often gave their commissions to larger shipyards in New York. These yards offered accommodations such as marine railways and longer launch cradles needed for the larger ships.

Along with the loss of commissions, David Williams also faced bankruptcy due to the loss of power on the river as more dams were constructed to power manufacturing companies and those companies grew. By 1840, the Hough-William's Dam from 1730, the Williams-Elisha Comstock dam of 1826, the Pratt-Clark-Williams dam of 1700, the Bull-Comstock-Griswold Dam and soon-to-be-added Mason H. Post Dam were all constructed about a quarter mile from North Main Street near the Williams yard. Don Malcarne wrote,

> *The demise of this operation came as the result of specialization and demand. The dam (or actually the amount of water backed up) could not supply enough power to operate so many businesses. Other dams sprung up that were specifically designed to power one business.*

Mason Post's dam was a perfect example of this. He purchased Ezra Williams's former comb shop and moved it up the river, above the Williams dam, where he built a dam to operate his ivory shop exclusively. Like most bankruptcies, a number of factors resulted in the demise of David Williams's businesses on the Falls River, but family support made the ordeal less painful.

David Williams lost not only his shipyard in the bankrupty but also his sawmill and his home. However, his younger brother Richard P. Williams stepped in to help settle the bankruptcy matters. Richard was able to allow David and his family to remain in their home by taking out a mortgage on the property for them. The Williams sawmill was taken over in 1848 when Gladwin & Wooster Company purchased the business. Six years later, it moved the sawmill from Meadow Woods to Novelty Lane.

The year of 1839–40 marked the end of an era for the Williams family, especially for David Williams, who lost his businesses initiated at the turn of the century by his father and a familiar place for all of the Williams brothers. By 1839, three of David's brothers had moved away and two of them were deceased. Their mother, Irena Williams, died on May 15, 1839, marking another tremendous loss for the remaining Williams siblings and grandchildren. But endings and transitions lead to new beginnings and changes. Richard P. Williams was the only son of Samuel Williams remaining in Essex who prospered. Richard and Louisa's son Ezra was born in 1839, a blessing and new beginning for the Williams family.

There was a silver lining in David Williams's dark cloud of bankruptcy. From 1835 until the early 1850s, Captain Frank West built several three-masted schooners at Williams Shipyard for the New York–based Lane, West & Company's Mobile routes. The company also managed a Havre line of packet ships. Captain West eventually moved to New York in the early 1850s and died there in 1856, but his use of the Williams Shipyard kept the yard in operation for more than a decade after bankruptcy. Since David and Frank were old friends and worked together at the Falls River yard for twenty years, Captain West probably employed David Williams to help build his three-masted schooners.

Shipbuilders, Shipmasters and Families

Olive Rockwell's eldest son, John Everest Rockwell, went to sea as a boy in the mid-1830s. Captain Rockwell was a shipmaster by the age of twenty-one and had a passion for sailing schooners. In 1840, John was master of schooner *Select*. At the age of twenty-five, he married Ruama Ayer from Saybrook, and the couple had a daughter the following year, named Olive after John's mother. Captain Rockwell eventually sailed packet ships in the transatlantic trade, substituting whenever cousins or uncles requested a break from their regular service.

In 1841, Captain Joseph Post sold his house on West Avenue to his brother David and decided to make the most of his retirement years. He and his wife, Mary, retired to the old Post homestead of his mother Lucretia Post's family in Deep River. The Denison-Post homestead property was located between Post Cove and Devil's Wharf and included farmland. The Posts had three sons by the time they moved—Prichard, David Rowland and Owen Lindsey—and six more children were born in Deep River. Two daughters named Mary and a son, John Prichard Francis, all died by the age of five, but Sarah Kate, Thomas Lindsey and Joseph lived to adulthood and married, as did their three older brothers.

After his retirement from the sea, Captain Post helped develop the town of Deep River and was part owner of several vessels, including early steamships. For a time, he owned Eustacia Island, a narrow island in the Connecticut River within view of his home, where he grew hay. As commissioner of highways for Essex and Deep River, Captain Post was

Photos of Captain Joseph Post and Mary Prichard Post, 1875. *Courtesy of R. Major.*

responsible for building several bridges and roads. He also was instrumental in getting the first lighthouse erected on the river. This was most likely the 1838 Lynde Point Lighthouse located on the west side of the entrance to the Connecticut River and Saybrook Harbor. A brownstone tower, fifty-five feet high, replaced an earlier, shorter wooden structure. A wooden spiral staircase ascended to the top of the tower, where ten lamps were lit and enhanced with reflectors. In 1852, the dim lamps were replaced with a fourth-order Fresnel lens that gave off considerably more light due to the reflection from its crystal prisms. Two years later, the lighthouse was given a fog bell. For twelve years, Joseph Post directed the Deep River Savings Bank, established in 1851. According to family records left by Robert Denison, Captain Post donated land to Deep River in order to create the Fountain Hill Cemetery, located near the Post Homestead, and he was named cemetery director, a position he passed down to his son Owen.

In 1841, David Rawson Post and his wife, Meriah, moved into the West Avenue home formerly owned by David's brother. With them were their three-year-old daughter, Lucretia, and one-year-old son, John Rawson, whom everyone called "Raws." It was a happy year for the family, as they were all

Captain David Rawson Post, oil portrait by F. Melzer, 1846. *Courtesy of R. Major.*

in good health and David was master of ship *Lorena* and gainfully employed by E.D. Hurlbut. He captained *Hector* for nearly six years, and it was time for a new ship. David approached the Hurlbut owners with the idea of having a ship built for use in the New York–Antwerp line. The company owners agreed, so David went to New City yard to talk over details with Richard P. Williams, master shipwright and shrewd businessman. Together, he and David worked out the ship's specifications in collaboration with the Hurlbut brothers. The packet would be 133 feet in length, 30.4 feet wide with a 21-foot draft. The owners decided the ship was to be named *Peter Hattrick* after Joseph Hurlbut's fifteen-year-old son, Peter Hattrick Hurlbut. The young man was ambitious and would eventually join his uncles' New York shipping company as a partner.

Like his cousin Russell Handy Post, whose ship *Elizabeth Denison* had a full-length maiden carved and painted for a figurehead, David Post's *Peter Hattrick* had a carved figure of a dapper young man, dressed in coat and tails, gracing the bow of the ship. If Richard P. Williams's packet ships *Elizabeth Denison* and *Peter Hattrick* were docked side by side, and their figureheads were not visible, it would be difficult to distinguish one from the other. Both vessels had similar lines, and their hulls were painted black with a thick stripe of burnt umber and three narrow white lines painted above and running the length of their hulls. The only way one could tell them apart was that *Peter Hattrick* was nine feet shorter. *Hector*, built in 1833, also by Richard Williams, had the same lines and design but no figurehead. Richard P. Williams's work was easily recognizable.

Regular shipmasters employed by New York–based shipping lines often commanded the same ship for a number of years. For example, John Urquhart was master of *Lorena* from 1832 until 1844. His younger brother Captain Walter Urquhart from Brooklyn, New York, took command of the ship on a substitute basis and while not in command of other ships. Grateful passengers who sailed to Mobile, Alabama, on *Lorena* gave

Ship *Peter Hattrick.* Reverse painting on glass by P. Weytz. *Courtesy of the Connecticut River Museum.*

Captain John Urquhart an elegant sterling silver engraved teapot, sugar bowl and creamer as a token of their admiration for the captain. The presentation tea set is now the property of the Wadsworth Atheneum in Hartford, Connecticut.

Trade items carried on *Lorena* included hundreds of tons of pig iron and coal, cases of merchandise, such as several hundred boxes of tin plates, bales of cotton, hundreds of firebricks, honey, horses and a couple of donkeys. Mercantile establishments sent in their orders to the shipping agents, usually located on or near the dock where the ships were berthed. Early on, when a ship was filled, it sailed. As more passengers purchased tickets, the ships sailed on more regular schedules. Many captains refused to sail on Sundays, and sailing on Fridays is still considered bad luck. All ships were at the mercy of weather conditions that could delay departures for several days. In one remarkable passage during the 1830s, Captain John Urquhart sailed *Lorena* from Lisbon, Portugal, to New York in twenty-five days. Westbound passages usually took longer, about thirty-five days, depending on tides, headwinds, season, weather conditions and route.

Ship *Peter Hattrick* (detail of figurehead). Reverse painting on glass by P. Weytz. *Courtesy of the Connecticut River Museum.*

Ships at that time generally traveled between 100 to 140 miles per day at 4.0 to 6.0 knots, or about 4.6 to nearly 7.0 miles per hour.

Captain Cornelius R. Doane from Essex was a regular captain who sailed in E.D. Hurlbut's Mobile Line. He started out with Joseph Post pioneering that line in the mid-1820s. During the 1830s, he finished a six-year run with the ship *Indiana*, followed by his new ship *Albamian*, built in New York, and commanded it for six years. Then he had a new ship built in East Haddam named *Cotton Planter*, which he sailed for several years before entering the shipping business for a few years in the 1840s. His last turn on the quarterdeck was in 1850 on the ship *Rhine*, which he gave over to his brother Captain William Doane in 1852. Their brother Captain Israel Doane captained packets in Griswold's Black X London Line and married Dolly Buckingham Post, Captains Charles and Russell Handy Post's sister.

The Doanes inherited the homestead of Ebenezer Williams, possibly a brother of Samuel Williams. Ebenezer purchased seven acres in Meadow Woods and built the house on what became Grove Street. In 1787, he married Deborah Doane, who passed the house down to her Doane relatives in 1852. Ebenezer's son Ebenezer Williams Jr. was also a house builder and probably learned his craft from his father. He built a house for his family right across the street from his father's homestead, where Captains William and Israel Doane and Dolly Post Doane lived.

Russell Handy Post and his wife, Roxanna, owned a considerable property in New City, including the parcel they lived on with their two young daughters, Frances J. and Martha. Captain Post walked out his back door to get to the shipyard and monitor the work being done, answer questions or respond to Richard Williams's requests or concerns. He could also check on sales at his store on the end of the dock, referred to as

Cap'n Handy's Dock. Captain Russell H. Post owned the entire store and workshop and three quarters of the dock. Captain Champlin owned the other quarter of the dock and several properties in New City, and both men co-owned New City Shipyard. In the 1840s, Captain R.H. Post also engaged in shipping and was captain of schooner *Martha Post*, which sailed out of New York. This schooner was probably named after Russell's sister Martha, who died at age fourteen in 1828, or his youngest daughter, Martha Post, who was thirteen years old in 1844.

Although it may appear that much attention was focused on the activity in the shipyards or the steamboat dock in 1845, there was a community rift simmering and drawing attention to the Old Point School building. The question was whether to repair the Old Point Schoolhouse near the site of today's Essex Savings Bank or build a new school. Several contentious meetings were held. The decision of what to do with the dilapidated school building was finally put to a vote. It was decided by a frugal majority of voters that the town would repair the old building. That could have been the end of the discussion, but some folks did not like the decision. Later that night, a loud explosion was heard coming from the direction of the old schoolhouse. Frantically, folks ran outside in their nightcaps to see what was going on. In their lamplight they were astonished to see all four sides of the building and the roof of the old school blown to bits by a couple kegs of gunpowder. The Baptist Meeting Hall on North Street was purchased and, for several years after, used as a school.

Miniature painting of Captain John E. Rockwell on ivory, 1840. *Courtesy of R. Major.*

Captain John E. Rockwell had a long career as a merchant mariner. Beginning with schooners used in the coastal cotton trade, he quickly transferred to transatlantic packets. While Captain Rockwell was at sea, his wife, Ruama, and daughter, five-year-old Olive, lived in a house built by George Harrington in 1815 on Pratt Street. Ruama and Olive also visited her parents, Thomas Ayer and Abigail Whittlesey Ayer, in Saybrook. On January 10, 1847, Ruama wrote a letter to her cousin Caroline Young in Farmington, Connecticut, about forty miles from Essex and ten miles west of Hartford. Ruama's letter allows us to travel back to Saybrook and Essex in the winter of 1847 and see what life was like for a young wife of sea captain who, like her older brothers John and Henry, was often out at sea.

> *Dear Cousins,*
>
> *It is so long since I have heard from you that I thought I would write and let you know that you still have a cousin at this place. I have inquired about you at every place that I thought it at all likely to hear, but the last news I can get is some 2 months old. I should be much gratified to receive a letter from some of my Farmington cousins. Thinking that perhaps you are a little like myself I will begin and give you all the news as far as I am able. My husband and 2 oldest brothers are both away—the first was to leave New Orleans for Liverpool England. My brother John is in Savannah catching the Shad. I wish I had a good one and could send you a part of it. Henry was at Mobile the last we heard from him and going to sail from there about New Years for Havre, France. he seems to be enjoying himself as he always is. He is in the ship* Martha Washington.
>
> *I have been spending 2 weeks at Essex on Pound Hill. did you ever go there Cousin Corny? I reckon so....I had a very pleasant visit. Ezra* [Clark] *goes up there to school* [Hills Academy] *so I could hear from home every day. Do you recollect going to the Lighthouse and seeing a young Lady there, she is married. She went away on a visit and came home and was published and going to be married Thanksgiving eve but no groom appearing she had to wait until saturday morning when she was married and went away and has not returned yet, report says he was an old bachelor and never was so far from home before in his life, perhaps she will get him out more often after this. I presume you have had all the particulars from Maria as she was there.*
>
> *How I do wish you would come and see us. I can show you my husband now if not in the original, something that looks like him. I suppose he thought I should forget how he looked so he sent me his portrait. it looks very much like him. Now don't you think you will come some of you. I don't think of any thing very interesting to write. We have had some fine sleighing but I could not improve it for want of a bearskin. Don't think that too bad—it is snowing now so I think we will have more....Ma says give my love to them all and should be glad to see them.*

Olive begins to read some and talks almost incessantly. How does Sophia get along with her kittens and babies. Now do come some of you. Make us a visit when it comes warm weather. Meanwhile let us hear from you often. There is so many of you that I should think you might fill a sheet often but poor me I have 3 absent ones to write to so often and onle [sic] *me to write that it takes me some considerable part of my time to keep track of the family. Give my best love to your Father Mother Brother & Sisters and let me hear from you very soon. Please excuse all mistakes and so forth as I am in haste and so many talking that I hardly know what I have written. Good bye dear Cousins.*
This from your Cousin Emma if you know her by that name. if not Ruama M. Rockwell

Perhaps the letter was never mailed or was returned, as it descended down the family for five generations with other fragments of the past. It was one of Ruama's last letters. John E. Rockwell wrote in his Bible, "Ruama Matson Rockwell died, December 10, 1847." Her mother died three weeks later, and mother and daughter are buried in the Upper Cemetery, also known as Junction Cemetery in Old Saybrook. Olive Post, five years old, remained in Essex with her father and his housekeeper, Mary French, when Captain Rockwell was out at sea.

David and Meriah Post had their fifth and last child on March 6, 1847, and they named him Lewis Walter Post. Meriah traced Lewey's newborn hand in order to make him some mittens. Numerous pinholes in the tiny hand pattern attest to the number of times it was used.

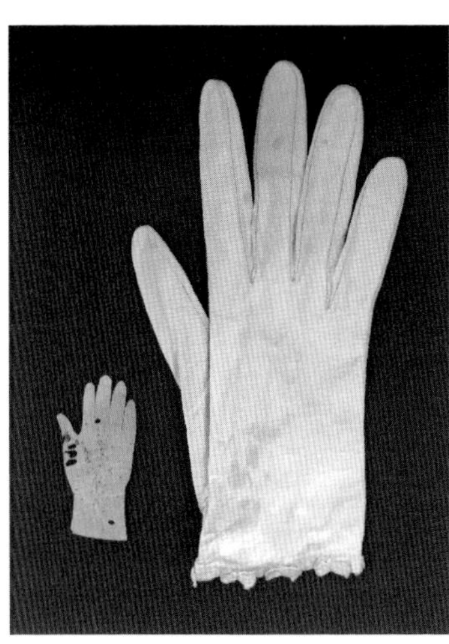

Lewis Post's newborn hand tracing, a pattern for his mittens. *Courtesy of R. Major.*

The year 1847 was a challenge for David and Meriah. Their new baby's arrival left Meriah in a weakened state of health, and David was needed at home. His brother Joseph came out of retirement to help, as did their cousin Captain Ezra Denison Post. Both men were masters of David's *Peter Hattrick* that year.

In 1848, John and Ann Urquhart's eldest daughter, Melvina Urquhart, then eighteen years old, and Captain Thomas L. Masson, aged twenty-three years, were married on September 7. Thomas had his sights set on becoming a shipmaster like his father and went to sea under Captain Edward C. Williams, who also built ships. In time, Captain Williams gave command of *Jane E. Williams* to Thomas Masson, and he remained master of the ship until 1853, when he transferred to the ship *Seine*, again formerly captained by Edward C. Williams.

Melvina Masson lived with her mother, Ann Urquhart, on Denison Road while her husband went to sea. She soon added to the Urquhart-Masson household and had help with her three wee ones from her mother and younger sisters.

Captain John Rockwell married Ruama's cousin Mary L. Ingham of Saybrook in 1848. Mary was the daughter of Asa Ingham and Mary Lord, and first cousin of Lydia Ingham, daughter of Judge Samuel Ingham. Lydia was the wife of Judge James Phelps, a student of her father and good friend of John Rockwell. Mary Ingham Rockwell treasured a red leather-bound autograph book. Most of the entries were religious in nature and emphasized the joys of life in heaven, but an acquaintance of Mary wrote the following entry about friendship on February 17, 1845, while she was visiting Harrisburg, Pennsylvania.

Friendship

Beauty with all its gaudy Shows
Is but a painted bubble.
Short is the triumph Wit bestows
Full of deceit and trouble.
Fame like a shadow flies away.
Titles and dignities decay.
Nothing but Friendship can display
Joys that are free from trouble.

In January 1849, Captain Rockwell took charge of *Peter Hattrick*. He sailed down to Apalachicola, Florida, with passengers and freight for E.D. Hurlbut. While he was away, Mary gave birth to a son on July 30 and named him John, but the infant died soon after his birth, "of fits." At the time, Captain Rockwell was in Antwerp, Belgium, with *Peter Hattrick*, loading freight and 255 passengers for New York.

Captain David Post contracted consumption or "wasting disease," which today is known as tuberculosis, and this may have been a reason

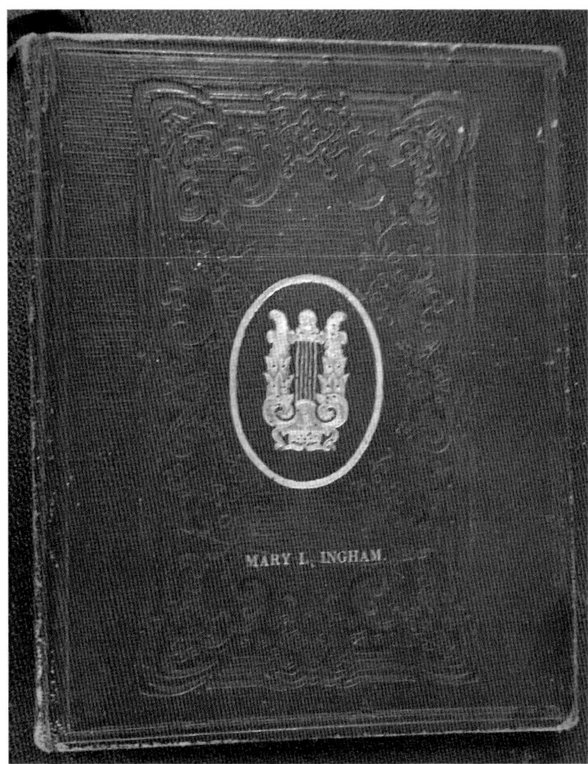

Mary L. Ingham's autograph book, circa 1840. *Courtesy of R. Major.*

why he was not captain of *Peter Hattrick* during 1848 and 1849. Joseph D. Post substituted for his brother in 1848 and sailed the ship in the Black Star Line to ports in England and Ireland. David Rawson Post died on November 4, 1849, leaving his wife, Meriah, then thirty-six years old, and four young children: Lucretia Maria, eleven; John Rawson, nine; Isabella Susanne, seven; and Lewey, two.

The extended family came together to help David Post's family. David's cousin Dolly Post Doane was very close to Meriah. To acknowledge their affection, Dolly and Israel named their first child Maria Urquhart Doane. No doubt, Dolly Doane and Ann Urquhart helped Meriah cope with David's funeral and the sorrow-filled months that followed. John Rockwell stepped in to help keep *Peter Hattrick* in service and was master of the ship from 1849 through 1851. The vessel was eventually sold to a Belgian company at New York in 1855 and converted to a man-of-war. *Peter Hattrick* was nearly fifteen years old.

In 1850, a few months after losing their son, Captain John and Mary Rockwell learned they were going to have another child. On October 30,

1850, Mary gave birth to a daughter, and they named her Mary Ingham Rockwell. The baby survived, but her mother grew weaker and died on December 29, 1850. Captain Rockwell continued working in E.D. Hurlbut's shipping line, as he had his household to support and wanted to help his cousin David Post's family also. At thirty-six years of age, John Rockwell began the new year of 1851 with his second wife Mary's funeral, a commitment to be a regular master in E.D. Hurlbut's Line and two young daughters, Olive, age seven, and Mary, two months. Widow Meriah Post and her daughters Lutie and Isabella helped John's housekeeper care for his daughters while their father was out at sea.

As was customary in the family, John E. Rockwell took his old Post family Bible to sea with him. The Bible was originally Captain David and Lucretia Post's, and they gave it to their daughter Olive, who sent it to sea with her son John when he went to sea at age fifteen. John always carried his grandparent's Bible to sea with him, but one day, he nearly lost it.

In the rush and turmoil of unloading and loading hundreds of passengers in Antwerp and New York, Captain Rockwell left his heirloom Bible behind in one of the ports he disembarked from. As providence would have it, a note, written by the person who returned the Bible was tucked within its pages for nearly four generations. It stated, "Captain John E. Rockwell, Care of Messrs. E.D. Hurlbut & Co. New York, or to Care of Jules Van Eeten, Ship Broker Antwerp, Belgium." The lost Bible was returned to Captain Rockwell, and the note traveled down the generations along with a bookmark with one important word embroidered in tiny stitches, "Mother."

Family events, losses, blessings and challenges experienced by the Hayden, Williams, Doane, Post, Urquhart and Rockwell families were typical of those experienced by other shipmasters and shipbuilders who resided with their families in boroughs and towns along the Connecticut River. Through their logs, record books, Bibles and letters, we get a glimpse of the day-to-day experiences and struggles of many maritime families. Although women and children are often only marginally included in books on shipbuilding, shipping and merchant sailing, they were and continue to be an integral and essential part of our story. Their contributions of time, gardening and herbology, childcare and education, housekeeping and sewing, home heating and cooking, nursing and midwifery, letter writing and recording of family history and genealogy, all accomplished in the absence of their husbands, is to be greatly commended and appreciated.

Launch of Ship Orphan

November 8, 1845

One of the most prolific Essex shipwrights of the period between 1844 and 1854 was Nehemiah Hayden, a master shipwright who brought the Hayden family back to New City. Nehemiah, born on March 29, 1819, was the son of Horace Hayden, grandson of Nehemiah who died in Bermuda, great-grandson of Uriah Hayden who built the *Cromwell* and second great-grandson of the Nehemiah who built a snow in his yard for the West Indies trade. In 1844, Nehemiah built the 397-ton bark *Jane E. Williams*, and the following year he built the ship *Orphan* at New City yard.

In 1844, Captain John Urquhart was master of the ship *Lorena* for the Hurlbut's Express Line, which sailed from New York to New Orleans. After coastal trading for over a decade on *Lorena*, Captain Urquhart was ready for a new ship and negotiated with shipping companies for the most lucrative deal. He gave the commission to New City owners Henry Champlin and Russell H. Post. Details were worked out, ship drawings executed with precision. A half-hull model was carved in order for the shipyard workers to see and know exactly where the timbers of the ship's hull should be positioned and what the shape of the hull would be. The ship was to be built under Richard Williams's supervision by shipwright Nehemiah Hayden. Richard was then forty-nine years old and Nehemiah twenty-six and probably taking on some commissions at New City for Richard. Another shipwright, J. Ellsworth, assisted Nehemiah with the building of Captain Urquhart's ship as is a common practice even today.

It was decided that the new ship would be 157 feet in length, 31.1 feet in width and have a 21.6-foot draft. The packet would have two decks and weigh about 682 tons. John Urquhart and his sons, John H., age twelve, and William Wallace, age seven, may have walked down from their home on Denison Road to observe the progress being made on the new ship. While their father was conferring with shipwrights, the boys could run down Handy's Dock to the store and look over the ships being finished off at the end of the pier. In time, John and William Urquhart would follow their father's footsteps and make their living on the sea.

In April 1845, Nehemiah Hayden and J. Ellsworth started work on *Orphan*, the name Captain Urquhart chose for his ship, perhaps due to his being orphaned at age thirteen. *Orphan*'s keel was hand-hewn and ceremoniously laid out on the ways with a brief laying-of-the-keel or "coin ceremony" initiated by Richard Williams and the shipwrights. This was customary and acknowledged the beginning or birth of the ship. A coin or two was placed under the keel block in order to bring the ship good luck. Then the date was duly noted in the yard's logbook as the start date for the shipbuilding project.

Adding frames to the keel was a hard task done by skilled joiners and a crucial step in the framework, like securing ribs to a spine. After months of planking and numerous other tasks, the yard crew celebrated with a toast when the shutter-plank, the last wooden plank to be added to the hull, was finally shaped and hammered into place. Members of the crew took turns hammering the last "golden spike," covered in gold leaf for luck, into the last section of the shutter-plank and toasted to Neptune and the ship's future. Then the work of caulking, sanding and tarring began. By the end of October, a date was set for the launch.

Launching Day was a community event that no one wanted to miss. November 8, 1845, was a day long remembered in Essex. Friends, neighbors and folks from Saybrook, Lyme, Haddam, Deep River and as far away as New York traveled to Essex to see the launch. Captain David Post was on his way to Antwerp, unable to attend the celebration, but Meriah Post arrived early and went to help Ann Urquhart with the ship's christening. Meriah's daughter Lutie and son Raws looked after their little sister, three-year-old Isabella. Ann Urquhart stood near the ship's bow with her daughters Melvina, Kate, Catherine and Sarah. Young John and William Urquhart arrived at the shipyard at dawn with their father, John, eager to watch all the last-minute launch preparations.

Joseph and Mary Post and their three older sons took the steamer down from Deep River in the company of the Pratts, Arnolds, Denisons and family from East Haddam. Others sailed up the river with the incoming tide or took a New York–Hartford steamer and stopped at the Essex Steamboat Dock, now home to the splendid Connecticut River Museum.

Captain and Mrs. Henry Champlin, Captain and Mrs. Russell Handy Post and a number of E.D. Hurlbut and Black Ball Shipping Company owners and their wives stood on a platform near the launch site with the minister and other dignitaries from the town. They were all dressed to the nines in the latest fashions. Captains William and Israel Doane and his wife, Dolly Buckingham Post, brought two baskets of food and libations, and several members of the Hovey, Pratt, Williams, Starkey and Stevens families also arrived with their baskets brimming. Everyone knew they would spend the afternoon with their families, friends and acquaintances and enjoy the festivities as they waited for the ship to be launched. The day was sunny and unseasonably warm, with a slight westerly breeze. Children tossed hoops and chased one another around the back of the yard.

The launch was to start at 4:00 p.m., when the tide was high, as was usually the case. Great gobs of cream-colored tallow were smeared on the marine railway so that the ship would slide easily down the ways. The great ship's hull was decorated with colorful flags and an evergreen bow hung from the bowsprit for safe voyages.

John Urquhart invited his rector from St. John's Episcopal Church, the Reverend Joseph Scott, to christen the ship. Reverend Scott was a Trinity College graduate who came to Essex from a pastorate in Derby, Connecticut. His tenure in Essex was to end after serving one year, and he planned to move on to Stratford, Naugatuck and North Haven. Dressed in his full-length black vestments, Reverend Scott headed toward the ship's bow. Everyone gathered around him and waited quietly.

A junior warden handed Reverend Scott his *Book of Common Prayer*. Captain Urquhart stood at the bow of the ship with his sons on either side of him, their heads bowed, while the rector read aloud in a deep baritone voice. "O, ye Seas and floods, O, ye whales and all that move in the waters, Bless ye the Lord. Praise Him and Magnify Him forever." Reverend Scott looked up at Captain Urquhart.

"What name shall be given to this ship, sir?"

"She shall be named *Orphan*!"

The rector sprinkled the ship's bow with holy water and continued:

> *Great God of Heaven and Earth and Sea, we praise you and thank you for all your blessings. To all who built this ship* Orphan, *and all who will*

> *sail and maintain her—May God's protection and guidance be with each and every one of you. Let us pray for the safety and wellbeing of all those involved in the life of this ship,* Orphan. *In the name of the Father, the Son and the Holy Ghost, Amen.*

As Reverend Scott finished his blessing, Ann Urquhart stepped forward and stood at the bow of the ship. Meriah Post handed her a bottle of spirits tucked into a tightly woven burlap sack, tied at the neck with a ribbon. With a nod from her husband, Mrs. Urquhart smashed the bottle against *Orphan*'s bow to the loud cheers of the crowd. The village brass band played a lively tune, and the children danced and clapped. A ship's carpenter secured chains to the sides of the rudder. The heavy chains were attached to lines drawn up on either side of the ship and tied to the capstan. This allowed the ship to be steered and the rudder to be directed where the captain wanted his unfinished ship to go.

When the tide was about to ebb, all preparations readied and onlookers ordered to stand a safe distance from the ship, Richard Williams and Nehemiah Hayden gave a signal to their crew lining the sides of the keel. All at once, their hammers banged wedges between the ship and the chocks, freeing the keel on both sides. The chocks were knocked away, as were the stops at the ends of the ship's cradle. Everyone held their breath waiting for the massive black packet ship to slide backward down the ways into the water. *Orphan* did not move. The crowd backed away several steps, unsure of what might happen. They prayed the launch cradle would keep the ship upright.

Orders were given to toss several hawsers over the bulwarks and heave on them, to pull with all their weight. The yard crew grabbed onto them and tugged the thick ropes in unison. They heaved on the hawsers to the point of exhaustion, and men slipped in the gravel and sand near the shoreline, but still, *Orphan* did not budge. More tallow was added to the ways. Every man on the banks of New City shore who had built a ship or gone to sea knew they were witnessing an ominous situation. They all feared the same thing, though no one spoke a word of it. They knew if a ship launched with difficulty, it was understood to be a dreadful omen. Bad luck would follow the ship.

After a brief discussion among the captain and shipwrights, Richard Williams called for all the oxen in town. Men and boys dashed home in all directions to yoke their oxen and grab their goads for prodding the animals. Soon, teams of oxen, secured in their yokes and chains, appeared on the cart path at the base of New City Street. More teams were driven down from Meadow Woods. It was not uncommon to call for oxen and their drovers during a launch when a ship was stuck on the ways. Ox teams, hitched together in trains of several beasts on either side of a vessel, could haul the ship back a few inches to allow for more tallow or forward to nudge the ship free of an obstruction and send it down the slippery ways.

Just as the hitchers were securing double-yoked ox teams to make the ox trains, the sky darkened rapidly from the northeast as storm clouds rolled in and the winds breezed up. Without warning, rain pelted the people. They tossed coats and blankets over their heads and dashed for cover. At the same time, drovers were given the order, "At the sound of the ship's bell, haul forward, one step toward the cove." Richard Williams rang the bell. "Giddup!" shouted the drovers, then quickly, "whoa!" to halt their oxen. The ship creaked and inched toward the cove. With one more well-executed tugs from the teams, *Orphan* started. The ox trains were immediately unhitched from the hawsers to spare the poor beasts a frightening ride into the cove with the ship. *Orphan* slid down the ways slowly at first but rapidly picked up speed and splashed into the water. Just as planned, the rudder held fast and *Orphan*'s stern veered north.

Launch of Ship Orphan, *November 8, 1845,* by R. Major. *Courtesy of R. Major.*

The rain let up. People cheered and shouted in delight. Men tossed their wet hats and caps into the air. Waves from *Orphan*'s launch rolled on shore, and sailboats rocked back and forth in the cove. Church bells pealed out from all the town steeples. Richard Williams, Nehemiah Hayden and J. Ellsworth shook hands while the yard crew filled their jugs with spirits and slapped one another on the back.

Orphan took on little water as its dry hull swelled, tightened and adjusted to being in the cold water, a tribute to masterful builders and caulkers who made all its seams between the planks watertight. The tar-covered oakum, or frayed rope fibers, caulkers skillfully hammered into the seams, in rhythmic loops and twists, left no cracks or crevices for water to enter.

A small steam tug hitched to a chain near *Orphan*'s bow and hauled the hull to the end of Handy's Dock, where it was moored securely to the dock pilings. Over the next three or four months, within feet of the Cap'n Handy's old workshop and store, *Orphan* was finished off and fitted out. Its three masts were stepped in on top of coins placed on the keel for luck. The ship would not be complete until its ornately carved and white painted billet head was added to the bow, iron hardware and fittings were installed, carpentry work was finished, the standing rigging and the complex network of running rigging with blocks and tackle was installed and approved and all the spars and sails were in place. *Orphan* was a fully rigged ship when all of these tasks were completed; its bright work was shined, interior finished carpentry was varnished and the name was artfully painted on its stern. Registering the new packet ship would be the next step in the process of getting *Orphan* underway.

Following a final blessing of the ship by Reverend Joseph Scott and a warm send-off from friends and family on a cold February morning in 1846, Captain John Urquhart set sail for his maiden voyage on *Orphan*. He departed Essex with his officers and crew of sixteen able seamen, knowing he might take on a few additional mariners and their sea chests at the Battery in New York. *Orphan* was moored a few docks east of the Hurlbut Dock at the Black Ball Line Dock on Packet Row. There, it was loaded with merchandise collected from merchants by C.H. Marshall, the shipping company that owned the Black Ball Line. Beginning in 1818, the line ran from New York to Liverpool, England, and back. Later, it expanded to include stops at Boston, Massachusetts, and Philadelphia, Pennsylvania. Once the freight, additional seamen and commercial passengers boarded, Captain Urquhart was bound for Liverpool, to unload merchandise for Manchester, Lancaster and surrounding counties.

Orphan reached Liverpool in early March 1846. It was scheduled to leave Merseyside around March 14 and arrive back in New York by late April. Exactly when John Urquhart contracted a serious illness is unknown, but he returned to New York and went home to his wife and six children in Essex. John's sister Meriah wrote in her leather-bound family Bible, "Captain John Urquhart died on July 17, 1846, age 40 years, 4 months." Some mariners believed the difficult launch of *Orphan* was a harbinger of John Urquhart's untimely death, after just one voyage on his new ship.

The death of John Urquhart left his wife, Ann, widowed, head of household and single parent to their six children. She had a housekeeper, as most wives of sea captains and successful merchants did. Some, like the Cheneys, had several staff to help with household chores, wood chopping, harvesting, cooking and other needs. In an effort to help Ann keep *Orphan* in service and income coming in, her relatives met to discuss who might be available among their circle of shipmasters to take over the ship. John E. Rockwell agreed to take immediate command of *Orphan*, but thirty-two-year-old Calvin C. Williams, son of Calvin and Eunice Williams, would become her regular master.

Captain Williams became a shipmaster by 1837, at the age of twenty-two, and with more than a decade of experience driving schooners and packets, he was prepared to take over as shipmaster of *Orphan.* He remained with the ship until 1854. According to historian Thomas A. Stevens, Captain Calvin Williams "made several notable runs" during his tenure on *Orphan.* In 1855, Captain G. H. Kempton ran *Orphan* in the Pelican Line from New York to Mobile, and later that year, he drove it to New Orleans. The following year, he sailed *Orphan* in the Old Line, which went from New York to Le Havre, France. Captain John E. Rockwell was master in 1858, and his third mate was his uncle Joseph Post's son Prichard. In 1863, *Orphan* was sold to a company in Germany.

Transitions

A period of transition for Captain Henry Champlin began in 1840. He remained with the Griswolds' Black X Line for his entire career at sea, from the mid-1820s when he initiated the company's New York to London line. During his later career, he commanded several beautiful packet ships, including *Cambria* in 1829, *Sovereign* in 1830, *President* in 1831, *Philadelphia* in 1832, *Montreal* in 1833, *Westminster* in 1835 and *Mediator* in 1836, the largest of all the Black X London Line packets at the time. For the next four years, until 1840, Captain Champlin was the senior captain of the Black X Line and superintendent of the line. He retired from the Black X Line in 1840 and lived with his wife, Amelia, in their stately house in Champlin Square, today lovingly maintained and decorated in period furnishings by current owner Geoffrey Paul.

Captain Champlin continued his co-ownership of New City Shipyard and oversight of several vessels for which he was the managing owner. As the decade came to a close, families experienced losses and gains, but overall, there was tremendous optimism about the future. This was due in part to the arrival of talented craftsmen, skilled factory workers and domestic help from European countries. Newer, bigger ships and luxury steamers presented advancements in commerce and transatlantic travel for business and pleasure. Opportunities abounded for developing mercantile enterprises utilizing new inventions and advancements in machinery, tools and technologies.

By 1852, Meriah Post had decided to move to a more modest home with her children. The sprawling West Avenue house felt like an albatross

Meriah (Urquhart) Post, oil portrait, 1850, artist undetermined, *Courtesy of Judy Rockwell Micoleau.*

with a six-foot wingspan on her shoulders. Even with the help of her Irish housekeeper Rebecca, managing the household and making repairs became too much of a burden. Close friends and relatives came together to help Meriah make the necessary move. Henry and Amelia Champlin managed the financial and logistical aspects and sale of the West Avenue house in a way that could have been beneficial to them.

Henry Champlin was the first president of the newly established Essex Savings Bank in 1851, and he handled Meriah's sale. The Champlins hoped to sell Meriah's house to their son-in-law Eben Stephenson, in the hopes that their daughter Eliza and grandchildren would live nearby. North Main Street was just being laid out at the time, and Amelia owned property on the north side of the new road that was part of her inheritance from her grandfather Ebenezer Hayden. Eben Stephenson purchased the house from Meriah, enabling her to have a new home built, but the situation did not work out for the Stephensons. Within two years, Eben mortgaged the house to Henry Champlin and returned to New York in 1854.

David Post's uncles and cousins came together to build Meriah's new house on what became North Main Street, and Meriah and her children remained in this house for about twenty-five years. Rawson Post became a ship captain and was often out at sea. While chief officer of *George Hurlbut*, Rawson sailed under his cousin Melvina Urquhart's husband, Captain Thomas L. Masson. In 1873, Rawson received a gold watch from a group of citizens of Wales for rescuing the entire crew of the barque *Olive* and risking his life in the process.

One family legend was that Isabella Post lived in the same boardinghouse as John Wilkes Booth in Washington, D.C. John's brother Edwin Booth, the popular, handsome and wildly talented Shakespearian actor, also lived for a time in D.C. Curiously, his photo card from 1893 was passed down with Post family photos. Most likely, Isabella was one of a multitude of adoring fans with a cherished photo of Edwin. Isabella was listed as living in Washington, D.C., in 1878 and working as an office clerk, but nothing more is known about her experiences there.

Lucretia Post was also listed as a resident of D.C., staying in a room near her sister's, but "Lutie" was probably there as a visitor. In her later years, she took care of Essex judge James Phelps and his wife, Lydia, with the understanding that she could remain in their home in retirement, with servants, until she passed on. Lutie was one of the history keepers who stayed in touch with Urquhart and Van Vorst cousins in New York. She left letters from her cousin Belle and lists of dates and times of her siblings' births written down by their grandfather Robert Adems of Johnstown, New York.

The first half of the 1850s marked a period of continued growth in passenger, immigrant and commercial activity for Essex shipmasters, but there were changes taking place in the shipyards and shipping companies. Immigrants from Liverpool, Scotland, Prussia, Ireland, Germany, Sweden and Belgium filled the steerage compartments of homeward-bound packets. Most immigrants wanted a better life for themselves and their families and viewed New York City as a beacon of hope and a destination to strive for, despite how treacherous and uncomfortable months in dank, cramped, steerage could be. Five million Germans and three and a half million British émigrés made the journey, and while most people welcomed the newcomers, one political party called the Know Nothings railed against the influx of Catholics and other immigrants.

In Essex, shipbuilding was carried on at the Williams yard until the mid-1850s under Frank West and at the Mack yard under David Mack until the early 1870s. At the New City yard and at Redfield and Parmelee yards, Nehemiah Hayden was still building ships. In 1849, Nehemiah built the

1,000-ton ship *Connecticut*, used in the transatlantic trade until the Civil War. The next year, Nehemiah built the 1080-ton ship *George Hurlbut* at New City yard for E.D. Hurlbut & Company. The ship was built under the watchful eye of Captain George West, the ship's captain, and launched from New City yard on November 5, 1850. *George Hurlbut* was highly praised by its owners. Captain West was master of the ship for only two years, as he died in 1852, but *George Hurlbut* went on to have a lengthy thirty-five-year career at sea under several Essex shipmasters including Thomas L. Masson.

In 1851, Nehemiah Hayden moved south on the river to the neighboring Redfield and Parmelee yard, which had a marine railway. There, on the western shore of the North Cove, opposite Braddock's Sail Loft, Nehemiah built and launched the 1,424-ton ship *Middlesex*. The ship was often described as the largest ship ever built in Essex. Showing his splendid talent as a shipwright, Nehemiah Hayden also built the magnificent clipper *Vulture* in 1853. It was launched on October 4 of that year. That same day, in the East Boston shipyard of another great master shipwright, Donald McKay, the 4,555-ton clipper ship *Great Republic* was launched and touted as the largest sailing ship afloat. Nehemiah Hayden's *Vulture* was built for the James Smith & Company of New York and was the only ship classified as a clipper built in Essex. In 1856, Captain Gurdon Smith was in command of

Ship *George Hurlbut*. Oil painting by A. Jacobson. *Courtesy of the Connecticut River Museum.*

Oil painting of Captain Elisha E. Morgan by Robert Leslie, R.A. 1847. *Courtesy of Robin Lloyd.*

Vulture and employed by the Post, Smith & Co. in its Regular Line. James Smith and Russell Handy Post were the owners of the shipping company.

Commissions for ships dwindled, and Frank West and Richard Williams were ready to retire. In 1855, Frank was sixty-five years old and Richard, fifty-nine. Shipping companies were purchasing packet ships from larger New York shipyards located closer to their offices. In 1854, Richard P. Williams sold his shares in New City Shipyard, but he continued to serve the town as a respected citizen and selectman.

Major changes took place during the early 1850s at two of the biggest shipping houses in New York. Griswold's Black X Line was transitioning to new leadership. Captain Elisha E. Morgan was a popular shipmaster. With his love of draughts, a popular board game, and his propensity for telling stories well, Captain Morgan was never at a loss for passengers who eagerly booked passage on his ships. Elisha Morgan was also greatly respected by his employer, John Griswold of the Black X Line. It was no surprise to anyone in the fleet that after taking his last voyage on the Black X line's new ship *Southampton* in 1852, Captain Morgan retired from the sea to become the next Black X Line supervisor.

Elisha Morgan made friends easily, especially on board his London liners, where the long ocean voyages to and from New York to Portsmouth and London afforded him the opportunity to make lasting friendships. Such was the case with several successful authors and artists, including oil painter Charles Leslie and author Charles Dickens. Leslie invited Captain Morgan to a meeting of the London Sketching Club, a group of about nine or ten painters of merit, including luminous landscape painter Joseph Turner. Captain Morgan was as comfortable entertaining cabin passengers on a London liner as he was having drinks and conversation with a group of London artists and writers. Within a few years, he was the sole owner of the shipping line, and he renamed the firm E.E. Morgan and Sons' Black X Line.

E.D. Hurlbut & Company also faced a number of changes in the mid-1850s. For over three decades, the firm was owned, managed and developed by brothers Elisha Denison Hurlbut, Samuel Hurlbut Jr., John Denison Hurlbut and George Hurlbut, all descendants of Thomas Hurlbut, a first settler and blacksmith of Saybrook and Wethersfield. Initially, most of their ships were built in Essex, but later, ships were also built in New York. Elisha Denison Hurlbut, born in 1801 in New London, Connecticut, was head of the shipping house. Under his leadership, the company grew to be one of the most prominent shipping companies in New York. Its office at 84 South Street, New York, was within walking distance of Packet Row, where all the packet ships were moored with their bowsprits reaching out over South Street toward the shipping offices.

George Hurlbut died in 1846. Samuel Hurlbut died in 1850, and head of the company Elisha Denison Hurlbut died on August 4, 1854. Only John Hurlbut was left to run the company. That same year, New City yard and store co-owner Captain Russell Handy Post and his partner John H. Ryerson began a business alliance in New York with E.D. Hurlbut & Company, handling their shipping. Post and Ryerson rented an office space from the Hurlbut offices and agreed to provide ships for the Hurlbuts' freight. The alliance faltered within weeks, as no shipping was given to Ryerson and Post, and John Ryerson joined forces with John Hurlbut. A Mr. Layton was added to the firm along with John's nephew Peter Hattrick Hurlbut, and the new firm was called Layton, Ryerson & Hurlbut Company.

Russell Handy Post severed business ties with John Ryerson, and on May 1, 1855, he opened his shipping firm next door, at 85 South Street, with W. James Smith and Benjamin Dunning. Post, Smith & Company became the New York shipping agents for Hurlbut's remaining lines, including the Rotterdam line begun in 1848, but Hurlbut closed that line in 1852. By the end of 1860, Post, Smith & Company took over ships to Antwerp for the Hurlbut Company and continued to keep that line in operation.

While several changes were taking place in the shipyards and shipping companies during the 1850s, ship captains were still plying their trade in the sturdy Essex and New York ships built in the 1840s and '50s. In 1852, Captain John E. Rockwell married Euphemia Noyes from New York. Euphemia died in 1854, leaving Captain Rockwell a widower again with two young daughters. Russell Handy Post came to his aid and offered John a position in his shipping company's Regular Line, which sailed out of New York. Russell Post assigned Captain Rockwell the one-thousand-ton ship *Frances B. Cutting*, a popular packet ship that carried hundreds of German and Belgian immigrants to New York from Antwerp for the Hurlbut Company.

In 1855, Captain Rockwell married for the fourth time, to Carolina A. Fisk from Hartford, Connecticut. He sold his Pratt Street house that year to Captain Jabez Pratt and moved his family to a larger house built for him in Essex. Constructed on land owned by Ethan Bushnell, the house was located on the hill overlooking Middle Cove. Captain Rockwell remained master of the *Frances B. Cutting* for three years, and worked for the Regular Line operated by Post, Smith & Company. Carolina and John Rockwell had a son in 1857 and named him John E. Rockwell Jr. John Sr. was forty-two years old. His life appeared to be going well. He had a comfortable home in Essex and was master of John Urquhart's ship *Orphan* in 1858; his wife and three children were healthy and all living under his roof.

In early June, tragedy struck again. After being ill for only two weeks, Carolina Fisk Rockwell died at the age of twenty-five on June 15, 1859. John Rockwell became a widower for the fourth time. With three children to support, he had no choice but to continue to work at sea. Post, Smith & Company took over the ships in Hurlbut's Antwerp Line in 1860 and continued to operate the line to Antwerp. That same year, John Rockwell took command of schooner *E.C. Howard*, built in 1854 in Essex and owned by Caleb Nickerson, C.H. Pratt and others. The 108-foot, nearly 309-ton schooner was launched in 1855 and sailed out of New York.

Captain Henry Champlin died in Essex on May 15, 1859, at the age of seventy-three. At the time of his death, he was president of the Essex Savings Bank and living with his wife in their home at Champlin Place. Retired captain Cornelius R. Doane took over Captain Champlin's position at the bank, and Horrace L. Pratt became the new co-owner of New City yard with Captain Russell H. Post. Henry Champlin left his wife, Amelia; children Charles and Elizabeth; and grandchildren. At the time, Amelia Hayden still owned considerable property, inherited from her grandfather in Essex, much of it in or near New City.

Joseph Post and Mary Post's son Prichard attended Hills Academy and went to sea under several Essex shipmasters, including Captain Henry R. Hovey, Amelia Hayden's younger half brother, on the Black X Line ship *Devonshire.* Prichard then shipped as third mate under Captain John E. Rockwell on ships *Peter Hattrick* and *Orphan.* He was also an officer on the ship *Francis Palmer* under Captain Russell Handy Post before being an officer under Captain Josephus Williams for a few years on the Boston ship *Berkshire.* By the fall of 1861, Captain Williams left the ship, and Prichard headed to Boston to take his first command. Following a career at sea that ended in the 1870s, Prichard married Eliza Sweet Williams, daughter of William H.

Williams of Essex, who left his house on River Road to Eliza. The couple resided there up until about 1920.

Prichard's brother David Rowland Post also became a shipmaster, but he went to a nautical training school in Liverpool on the advice of his maternal uncle. David spent most of his career commanding ships owned by large shipping firms in England and circumnavigated the globe twice during his career, which ended in retirement in 1899. David Rowland Post married Kate Loomis in 1868, and she and their children accompanied him on nearly all of his voyages. In his later years, he was postmaster for the town. Prichard and David's sister Cate Post married Captain George Morrison of Deep River, master of several famous Down Easters and a pioneer of steamships in the American–Hawaiian Line.

Captain Walter Urquhart, brother of Captain John Urquhart and Meriah Urquhart Post, had a dual career as a shipmaster and an inventor. His primary concern was for marine safety. In 1858, Walter began work on an innovative design for a raft that could provide a means of both saving lives and navigating when ships were sinking and rescue was not readily available. Captain Urquhart's life raft was made of nineteen mattresses filled with cork shavings and covered with an outer layer made waterproof with a substance from tropical trees called gutta-percha. The mattresses could be filled with air for more buoyancy and comfort, both in the water and when used on board ships for passengers and crew to sleep on.

In the event of a collision or other disaster causing a ship to sink or forcing passengers to abandon ship, the mattresses could be tossed overboard and easily and securely fastened together with belts and buckles in a brick-like fashion. Each mattress could support about 550 pounds, or two to four people, and the entire raft could be sloop or cutter-rigged to navigate with masts and sails taken from a longboat and tossed overboard near the raft. The mast was held in place by planks running the length of the raft at the center. Captain Walter Urquhart successfully experimented with his life raft in New York, the English Channel and the River Seine in France and won awards for his invention from various scientific academies. He patented his invention in every country and hoped to make his fortune and save lives through the manufacture of his lifesaving invention.

Other important inventions relied on the natural resin product gutta-percha. In the late 1850s and for decades thereafter, it was the material most effective in waterproofing underwater cables. The pliable but strong substance was sold by the Gutta Percha Company in England as early as 1845. By 1851, the company was producing flexible underwater cable tubes from this tree product to insulate the first cross-channel cables. The company went further to create the insulating tubes for transatlantic cables, first laid across the Atlantic Ocean in 1858, and it refined its manufacturing processes based on outcomes of each under-sea cable attempt. Gutta-percha was also used to waterproof sailors' sou'westers and other rain gear. Water streamed off the rain hats, and the resin coating was not adversely affected by salinity. Lighter than water, the product was perfect for covering rafts and life buoys. In fact, every part of the gutta-percha tree was used for some useful purpose, including flowers that were eaten and medicines made from parts of the tree for curing wounds. Gutta-percha is superior to India rubber, as it does not become brittle.

The development of the telegraph machine by Samuel B. Morse in the 1830s and his Morse code for writing messages resulted in the inventor sending the first message in the May 1840. This greatly facilitated communication. Within a decade, urgent messages were sent all over the country, as the new telegraph system was set up from the East Coast to the West. Homing pigeons, local post riders and even the Pony Express of 1860 quickly became obsolete.

By the mid-1860s, messages were being successfully sent through underwater cables from the United States to Great Britain. Telegraph communication not only helped open up the western frontier, but it also

Souvenir items made from 1858 transatlantic cable. *Courtesy of R. Major.*

greatly reduced the time required for messages to reach European nations and for them to communicate with the United States. By 1864, after several attempts, the transatlantic submarine cable successfully connected people on both sides of the Atlantic. The cable was another tremendous leap in technology, made possible by the natural insulating qualities of gutta-percha. The product was manufactured in Seymour, Connecticut.

By 1860, the wooden shipbuilding industry in Essex, Connecticut, had foundered. Don Malcarne wrote,

> *Ominous signs were on the horizon. Steam powered ships, soon followed by ironclad vessels, were coming of age. The great cities of Brooklyn, Boston, Baltimore, and Philadelphia were producing these ships, and by 1860, Essex Village was sinking into a state of financial and even psychological depression. An area built on a narrow economic culture was failing, as fewer and fewer sailing ships were constructed.*

The wooden shipbuilding decline was one factor that deeply affected the citizens of Essex, but perhaps the state of the nation, approaching a national crisis, contributed far more to the general malaise and feelings of trepidation.

Civil War

Abraham Lincoln was nominated for president of the United States in 1860, and soon after, South Carolina seceded from the Union. Several other Southern states followed suit. In all, eleven states, including Virginia and North Carolina, joined the Confederate States of America. On April 12, 1861, the Confederates fired on Fort Sumter in Charleston Harbor, and the American Civil War was underway. With the country in turmoil and young men called to arms, Essex families had serious personal matters to contend with, like seeing their young husbands, fathers and sons go off to war.

Captain John E. Rockwell knew exactly what he needed to do. He immediately made arrangements for the care of his children. John E. Jr., age four, was sent to live with relatives in Hartford, as was Olive Post Rockwell, who was nineteen in 1861. Olive met a young man named James Woodbridge Hale, a dry goods merchant from Glastonbury, near Hartford. James took care of his mother and aunts and managed the family home and store, while his elder brother Fred, a surgeon, and younger brother Cornelius, a surgeon's assistant, volunteered to serve the Union. Fred served on land, while Cornelius volunteered to assist a surgeon at sea. Mary Ingham Rockwell, age eleven, remained in Essex with her aunt Meriah and cousins. Rawson Post continued in merchant service.

Captain John E. Rockwell was the first Essex shipmaster to volunteer for service in the Union navy. At the time, June 1861, the navy was in the process of increasing its ships and personnel, developing plans to respond to

Confederate privateers and raiders and arm merchant vessels. The Brooklyn Shipyard was deluged with business, and from the beginning of the war to the end of the conflict in 1865, the navy grew to nearly ten times its original size. Credit for this rapid expansion was given to Secretary of the Navy Gideon Wells from Glastonbury, Connecticut. Secretary Wells was against the Union blockade of Southern ports, but he agreed to cut off the cotton trade that provided income for the Confederates to procure goods and war supplies. A series of blockades along the eastern coastline was formed to impede access to Southern ports and inhibit trade by the Confederacy.

England supported the Confederate states, as it was the biggest importer of American cotton, and English mills needed the product to remain in operation. Confederates greatly benefited from British-built blockade runners, like CSS *Banshee*, constructed in British shipyards. Naturally, tensions mounted between the United States and Great Britain over the issue of British-built blockade runners.

Captain John Rockwell was a merchant master until he volunteered to serve the U.S. Navy. He and other volunteer shipmasters were referred to as "acting masters." They were temporary officers under command of the Union navy but were not career naval officers and would return to civilian life, merchant shipping and immigrant and commercial transport after the war was over. Familiar with the ships, rivers and coastal conditions from working in the cotton trade, many merchant masters like Captain Rockwell made valuable contributions to the Union cause. On April 15, 1861, President Lincoln called for seventy-five thousand volunteers to serve for three months. Captain John E. Rockwell went to New York to volunteer his services, and on June 10, 1861, he was officially appointed an acting master.

Captain Rockwell first served on the USS *Wabash*, as did dozens of other acting masters. They were trained on the ship and transported to the Chesapeake Bay, where they received further orders and assignments. The 4,808-ton screw frigate *Wabash* was one of the most formidable warships in the early years of the Civil War. It was the flagship of the Atlantic Blockading Squadron and based at Port Royal, South Carolina. In early 1862, Hartford native Cornelius Hale, aged twenty-six, wrote to his older brother Jim, a close friend of Olive Rockwell, while stationed aboard the U.S. steamer *Isaac Smith*. Cornelius volunteered in the navy as a surgeon's steward, having had some experience working in an apothecary. After only a few months at sea, he became the vessel's physician, as the following portion of the four-page letter to his brother explains.

St. Helena Sound, Jan. 21, 1862
US Steamer Isaac Smith

Dear Brother,

Yours of the 3rd was received yesterday and you being the only one of the family who seems to think enough of me to answer my letter, I concluded I would try and write again, that I might know something of what was going on at home....

I do not know Mr. Rockwell of the Wabash, *but I am glad to hear you have found a friend in his daughter, and when we visit Port Royal again, if I get time to go on board the* Wabash, *I will have the gentleman pointed out to me, but unless you wish it, will not speak a good word for you, as you know that no one is allowed to talk with the men on board a man of war, without permission of the officer of the deck, however, I wish you success and don't imagine you are committing a crime while engaging the leisure hour of the young lady who seems to be pleased with your company. I wish some of these beautiful warm moonlight nights that I could spend the evening along side some pretty young lady...and when you are caught with the Blues, don't cherish them as you would a pleasant thought, but go out, take a glass of Lager, when you take it remember me, and if one don't do it, take another and think of Miss Rockwell—We have just got orders to up anchor at daylight, bound for Savannah, all the Gun boats, Frigates etc are ordered before Fort Pulaski. I have not had time to say half I want to you, now maybe by the time I get this in the mail, the battle will be over, it will undoubtedly be a terrible one, our boats will many of them be cut to pieces, this will be no such pay as the affair at Port Royal, great are the preparations we have made and our men are thoroughly drilled at the guns, night and day—sometimes in the middle of the night we hear the call to quarters, then all hands, officers and men and obliged to turn out of their hammocks, lash them up, carry them on the quarter deck, cast loose the guns, sponge, load and fire—the last time they tried this only six minutes from the call to quarters (at 1AM) the first gun was fired, quick work—we have quite a number of expeditions in small boats up the rivers leading into the Sound, one from this craft got back last week, they had a good time but the rebels held out of sight, the sailor on show is pretty well shown in the notes on the trip, here are a few—they camped on Edisto Island one night, officer in command ordered a small fire at night, so as not to attract the rebels, sailors went to work and set fire to a* [servant] *house on the plantation, making such a small fire that the sailors had to turn to, and help put it out, then they got another fire, and after supper, they got a wagon, put one sailor in, rode him all around, finally drawed him into the fire, burnt up the wagon, came near burning the sailor too, finally they were ordered to bed, they all turned into a cotton house, where they kept singing until 2 or 3 in the morning, when all was quiet, only now and then you could hear some expression such as this,*

"I say Bill, Bill,"

"What the hell do you want?"

"I say get up Bill and untie that damn dog loose that's howling outside that door."

The Doctor has been engaged on board the Sloop of War Dale *where he was ordered by the Fleet Surgeon to take the place of their Surgeon who was sent home on account of losing his eye sight, since which time, we have been up the river a week at a time absent—I have been Surgeon of the* [Isaac] Smith, *have had plenty to do, the Captain was very sick at one time—I had the luck to care for him entirely, this did not hurt my reputation you can bet. There is a good deal of sickness here and will be much more as soon as the hot weather comes on, which will be soon now, but I must close this letter now, and next time will try and get my letter begun before I get to the last page, tell Mother to consider all my letters as if written to her. Remember me to all, write soon and often & oblige.*

Your Aff brother
Cornelius S. Hale

Please direct your letters Care of Flag Ship Wabash *instead of* Susquehanna

By May 1862, Captain Rockwell was given command of the armed schooner *Hope* and assigned to the South Atlantic Blockading Squadron off Florida. *Hope* was a dispatch and supply schooner responsible for bringing mail and supplies to other vessels in the fleet. Since John Rockwell had a passion for sailing and racing schooners, he probably looked forward to the challenge of being part of the 1862 blockade off Fernandina Beach near Jacksonville, Florida. Captain Rockwell remained the acting master of *Hope* until April 1864 and spent the early part of the year as a blockader off Charleston, South Carolina.

During the Civil War, many Essex shipmasters were miles from home, either serving in the U.S. Navy as acting masters or continuing commerce and passenger service as merchant masters. Captain Charles Post was fifty-three in 1862. He was a brother of Essex shipmasters William and Russell H. Post. Charles Post was master on ships *Finland* and *Martha Washington* and brig *Winthrop*. In 1859, he took command of *Elizabeth Denison*, primarily owned by his brother Russell, as was the schooner *Handy*, both Essex-built coastal schooners. These commands were followed by schooners *J.S. Lane* and *Martha Post*. In 1862, accompanied by his wife and son Colton, Captain Charles Post sailed for Yates & Porterfield on the ship *Elizabeth (Denison?)* to the West Coast of Africa. Both Captain Post and his wife became ill with fever, probably yellow fever, transmitted by a mosquito, which in its toxic phase may cause delirium, seizures, organ failure, jaundice, coma and death. Captain and Mrs. Post both died of the fever, but their son Colton survived and returned home to Essex on the ship.

In March 1862, Captain Walter Urquhart was in Cuba on a trading venture, perhaps to buy more cork for his rafts, as there were native cork palm trees growing on the western part of the island. Cuba was a place he had sailed to many times over the years, beginning with his command of his brother John's ship *Lorena* in the mid-1840s. He was well acquainted with the shipping agents in Cuba and the weather and tides there. While contracting business in Cuba, Captain Urquhart fell ill. On March 6, 1862, he died of congestion of the brain, a cause of death often given to shipmasters in the nineteenth century, meaning brain swelling, possibly caused by a stroke, infection, concussion or a fall. The crew brought Captain Urquhart's body to Liverpool, where his remains were cared for at the Sailor's Home until the family was notified and preparations made for his return home for burial in Brooklyn. Captain Walter Urquhart was fifty years old. He was never to realize his dream of manufacturing and patenting his navigable life rafts for ships.

John H. Urquhart and his brother William W. Urquhart, sons of Captain John and Ann, continued in merchant trading and immigrant transport during the Civil War. John went to sea under his brother-in-law Captain Thomas Masson, master of the bark *Jane E. Williams* and took over command of the ship in 1855. John was also master of the 1,200-ton ship *Francis P. Sage* and brig *George H. Chase*. After the Civil War, Captain Urquhart was master of a side-wheel steamer, formerly the blockade-runner CSS *Banshee* that made several successful runs through the blockade from Wilmington, Delaware, to Cuba or the Bahamas, before the U.S. Navy seized it in November 1863. The vessel was recommissioned in June 1864 as the USS *Banshee* and served until the end of the war in the North Atlantic Blockading Squadron and the Potomac Flotilla. USS *Banshee* was decommissioned in November 1865 and sold. By 1867, it had been renamed *Irene*, and Captain John H. Urquhart became master of the steamer for several years and traded between the United States and the Bahamas.

John Urquhart's younger brother Captain William W. Urquhart attended Hills Academy and went directly to sea, as his brother John had done. By 1861, William was chief officer under his mentor, Essex captain Henry R. Hovey, a popular Black X Line shipmaster. Captain Hovey was master of the largest transatlantic packet in the line, 1,717-ton ship *Amazon*. The ship unfortunately burned and sank off the coast of England in 1863. After one year as *Amazon*'s chief officer under Captain Hovey, William W. Urquhart became a shipmaster in November 1861 of another Black X ship, *American Eagle*. As with other vessels in E.E. Morgan and Son's Black X Line, *American Eagle* sailed between New York and London, with stops at Portsmouth and occasionally other ports along the southeast coast of England.

Captain William Urquhart was twenty-three years old when he took charge of *American Eagle*. He enlisted mates from Essex to go with him on his first voyage as a new Essex shipmaster. One of William's friends and neighbors was Gilbert Williams, whose father, Ebenezer Williams, was a ship and house builder in Essex. Gilbert, age twenty-six, was well on his way to becoming a chief officer, with the goal of following his older brother Ebenezer Jr. and becoming a shipmaster. Christopher Columbus Doane was another good friend of William Urquhart. With a few years as an able seaman under his belt, William asked Christopher to sign on as second mate. The only surviving son of Essex Savings Bank president Cornelius R. Doane, a renowned Essex shipmaster, Christopher was twenty-two years old and ready for a promotion and adventure. William Urquhart also enlisted the help of four ship's boys from Essex, one of them being his cousin Lewey Post, David and Meriah Post's youngest son. As William Urquhart took his first command of *American Eagle*, Prichard Post took over the bridge of the *Berkshire*. The next generation of Essex and Deep River masters was set to follow in their fathers' footsteps and make their living as merchant mariners for the next decade or two. In William's case, he became master of several packet ships and did not retire from the sea for twenty more years.

When Captain Urquhart left New York on *American Eagle* in November 1861, he assured his aunt Meriah Post that he would take good care of Lewey. At the time, William had no way of knowing how perilous the journey would be or how his promise to Meriah would be tested. Years later, he would tell others that it was a disastrous voyage. Merchant trade continued during the Civil War, but the number of immigrant and cabin passengers dwindled.

Oil portrait of Captain William W. Urquhart by R. Major. *Courtesy of R. Major.*

A ship that could carry five hundred immigrants, such as *American Eagle*, found only a few dozen souls willing to brave the Atlantic crossings, which included a new threat. Confederate raiders could overpower sailing vessels with their steam engines, and they presented an ever-present threat. While passengers were not usually harmed, they could lose their jewels, money and possessions and endure the inconvenience of having the ship taken. The

Atlantic environment for merchant sailing ships was much more adverse than it was for merchant captains in earlier times. The Civil War hostilities and increasing numbers of steam-powered and ironclad vessels were the death knell for the Age of Sail. Masters of traditional wooden sailing vessels like packet ships and clippers knew their beautiful ships and traditional trading practices were coming to an end. Young shipmasters like Prichard Post and William Urquhart, heading overseas during the Civil War, faced perils their fathers and grandfathers never imagined.

Captain Urquhart's outbound winter voyage across the Atlantic was plagued with winter storms, heavy seas and violent gales. Sailors working aloft nearly froze their fingers to the rigging, and the ratlines were slick, glare ice. When Captain Urquhart finally reached London, he and his mate Gilbert found the town covered in black mourning drapes. Queen Victoria's husband, Prince Albert, died on December 14, 1861, to the country's dismay and the queen's devastation.

In need of new sails, Captain Urquhart and Gilbert Williams must have spent time at Hart's Sail Loft, located near the docks. John Hart was a retired sailmaker and owner of the sail loft and chandlery. He and his wife, Harriet, had three daughters: Emma, age sixteen, and her two half sisters, Helen Mary, about twenty, and Harriet, age twenty-four. The Hart sisters became close friends with William and Gilbert. It is most likely new sails and other necessities were purchased at Hart's chandlery and loft.

On the return voyage, the *American Eagle* was met with a violent gale off the rocky southwestern coast of Ireland, leaving all on board in peril. The area is well known for shipwrecks, and mariners have feared the rocky coast for centuries. As night fell, the gale raged on, tossing the ship about like a toy boat. Captain Urquhart prepared to ride the storm out, but the winds grew stronger and the ship swung around and broached to, leaving *American Eagle* on beam ends, with its masts parallel to the sea. Visibility was all but impossible, and half the instruments washed overboard. The fifteen-year-old ship was battered, strained and leaking profusely. Passengers in steerage cried out in terror, and the crew did all they could to right the *Eagle* but feared the ship would be swamped. Everyone knew death was imminent if the ship was dashed on the great rocks.

Captain Urquhart exhausted all measures of preserving the ship. At the mercy of God, and in the depths of anguish, he made his way to his large King James Bible, opened it at random and ran his fingers down a page until they stopped at Psalm 107. Trusting that God was speaking to him, the captain read aloud in the dim whale oil lamplight, swaying side to side.

Psalm 107, verses 23–31

23. They that go down to the sea in ships, that do business in the great waters:
24. These see the great works of the LORD, and his wonders in the deep.
25. For He commandeth, and raiseth the stormy wind, which lifteth up the waves thereof.
26. They mount up to the heaven, they go down to the depths: their soul is melted because of trouble.
27. They reel to and fro, and stagger like a drunken man, and are at their wit's end.
28. Then they cry out unto the LORD in their trouble, and he bringeth them out of their distress.
29. He maketh the storm a calm, so that the waves thereof are still.
30. Then are they glad because they be quiet; so He bringeth them unto their desired haven.
31. Oh, that men would praise the LORD for His goodness, and for His wonderful works.

Captain Urquhart was greatly relieved and trusted that the ship would be spared. Many times in his life, he used this method of opening his Bible at random and letting his fingers slide down the page to a passage. He once revealed that God spoke to him through this process, and he had absolute faith that his prayers for survival were heard. Some folks called Captain Urquhart a Bible seer. Others called him psychic, but he believed and trusted that the Lord spoke to him through the Bible and would answer his prayers and tell him what was to be.

The storm did not let up until morning, but the captain reassured his officers, passengers in steerage and crew they would survive. In the light of day, the condition of the ship and passengers was assessed. Aside from a sprained wrist and bruises, no one was seriously injured, but *American Eagle* was in no condition to make the homeward passage across the Atlantic. Captain Urquhart ordered the crew to man the pumps around the clock and jury rig the ship. When the winds died down and the ship was fully righted, they headed to Queenstown, now Cobh, Ireland, where the ship was put into dry dock. *American Eagle* was deemed unseaworthy and required extensive repairs. Captain Urquhart knew they would have to layover in Ireland until the repairs were made. He feared Captain Morgan would fire him from the line, but his employer was an understanding man and grateful that no lives were lost.

While waiting for his ship to be repaired, Captain Urquhart learned of a new ship being fitted out in Liverpool. It was said to be a merchant ship bound for Palermo, but word on the docks was that it was a Confederate raider. Hoping to avoid any run-ins, Captain Urquhart departed as soon as *American Eagle* was returned to service.

On the homeward passage, the captain received unwelcome news from a passing ship. His uncle Walter Urquhart died in Cuba while on a trading venture while buying cork for his rafts. William wondered whether his uncle's remains were in Liverpool when he was having his ship repaired right across the Irish Sea. It was days before he accepted the fact that his uncle's dreams of manufacturing life rafts would never be realized.

Captain William decided to take the southern route home and hoped to make up for some lost time by heading into the Gulf Stream. The ship handled like a new vessel with all the repairs and made excellent time for several days. When the ship was about mid-ocean, the unthinkable happened. The winds breezed up and the seas swelled.

Captain Urquhart looked up to see his young cousin Lewey Post scrambling aloft to free a line. He reached out just as the captain hollered for him to come down. The ship listed, and Lewey was tossed into the boisterous sea. The captain shouted, "Man overboard!" as he unlashed a chicken coop and tossed it overboard. Second Officer Christopher Doane headed for the longboat. As the crew prepared to lower it over the side, Christopher jumped into the sea to keep the boat from crashing against the side of the ship and being smashed to pieces. Captain Urquhart ordered his chief officer to take charge of the ship while he went up in the crow's nest to get a better view and direct the rescue boat. By the time he reached the top, he could not see any sign of Lewey, but when the ship was on the crest of a wave, he was able to make out the coop, end-up and drifting far away from the ship. Distressed beyond measure but trying to keep his wits about him, Captain Urquhart knew he had to find and save Lewey Post, but the boy had drifted a mile from the ship.

Captain Urquhart waved the rescue boat onward in the direction of the chicken coop, knowing Lewey would be somewhere near it, as it was only seconds between his cousin falling from the spar and the coop being tossed in. He ordered the ensign to be dipped, believing that if Lewey were still alive, he would notice the flag salute and know help was on its way. He might hang on a little longer, but the water was very cold. Captain Urquhart was consumed with horror, recalling his promise to his aunt Meriah. He would never be forgiven if he returned home without her son. Everyone aboard loved Lewey, not only because he was the youngest of the ship's boys but also because he was one of the most congenial and entertaining. Several of the crew fought back tears, and there was hardly a dry eye aboard ship. Then Captain Urquhart saw through his glass that the rescue boat was returning. He feared that his cousin was lost, and with the immense waves, he might lose his crew also. Instead of losing one, he could lose all seven. The thought of losing them horrified him.

As the crew rowed closer, the captain could make out two or three men rowing and one bailing. Then he saw four men rowing, and he was glad he sent an extra man to bail. Finally, his heart jumped for joy, as he saw six men rowing and one bailing, and he knew that Lewey was saved. The captain called down to his crew to let them know the good news. Everyone cheered, shouted and clapped with great joy and relief.

When all those in the rescue boat were safe aboard the ship, Captain Urquhart asked Lewey how he managed to stay afloat so long in the cold water. Before he could answer, the other ship boys revealed that they had made a pact at the beginning of the voyage, that if any of them fell overboard, the others would toss in the life buoys. Lewey said he was able to swim to one of the buoys, hold fast to it and then prayed as he drifted swiftly out of sight. He saw the chicken coop end-up and knew Captain Urquhart was sending the rescue boat, and when he saw the ensign dip, he said his heart leaped for joy because he knew the boat was surely heading his way.

Christopher Doane was a hero who risked his life to launch the ship's boat and row through high seas to rescue Lewey. When asked how he kept the boy from passing out from exposure and cold, Christopher told the captain that when they hauled Lewey into the boat, the boy lay on the on the floorboards shivering like a scup in a bucket. Christopher ordered him to get up and row to keep his circulation moving.

It was times like that, when ship's boys and crew came together to reach a seemingly impossible goal, that mariners were able to perform their greatest feats at sea. Together they rowed out on that treacherous sea in a small boat and labored to locate and rescue a young shipmate who had drifted a mile. In doing so, the mariners demonstrated their courage, quick thinking, camaraderie, strength and determination. No doubt, everyone aboard *American Eagle* that night celebrated Lewey's good fortune and their highly successful rescue mission.

In the spring of 1863, Captain William Urquhart married seventeen-year-old Emma Hart, John and Harriet Hart's daughter, from Mile End, London. The wedding took place in London. The following year, Gilbert

Oil Painting of ship *American Eagle*, after Samuel Walters, by R. Major. *Courtesy of R. Major.*

Williams married Emma's older half sister Harriet Gubb, whose father, Captain William Gubb, died when she and Helen were young. Both couples made their home in Essex, Connecticut, and raised their children there. In 1869, Helen Gubb married Captain Allen J. Griffing from Lyme, who also sailed for the Black X Line, and their son William was born in Essex on March 6, 1874.

Gilbert and Harriet Williams lived in his father Ebenezer's home and had a son and a daughter. Gilbert Williams became a shipmaster of several packet ships, including *Villa Franca* in 1868 for E.E. Morgan's Black X Line, *Plymouth Rock*, *Brazos* and *Sunrise*, a ship that sailed in the Far East trade to China and India. Emma Urquhart went to sea several times over the course of her husband William's lengthy career. Emma and William had a daughter, Kate, and a son, William Jr., in Essex, but eventually moved to Brooklyn, New York.

Second Mate Christopher Columbus Doane died young, in 1870 at the age of thirty-one, and is buried at River View Cemetery with his brothers, parents and other members of his family. Lewey Post went to sea for several years with William Urquhart and worked his way up to chief officer. The cousins went their separate ways in late 1867 due to the changing of ships.

Essex shipmaster Captain Isaiah Pratt saw his packet ship launch from a New York shipyard during the Civil War in 1863. Captain Pratt, one of the most well-known shipmasters in Essex, was forty-nine years old when he took command of the ship *Hudson II* for one relief voyage. Captain Pratt was principal owner of the ship but not its regular master, although he was master of several Black X Line ships during his career, including *Mediator*, *Hendrick Hudson*, *Margaret Evans* and *Southampton*. In his retirement, Captain Pratt was active in civic affairs and donated funds to build and operate Pratt High School, now the Essex Town Hall.

In late January 1863, while still in command of schooner *Hope*, Captain John Rockwell and his crew captured the schooner *Emma Tuttle*, just off Charleston, South Carolina. About the same time, President Abraham Lincoln signed the Emancipation Proclamation, freeing all slaves in the Confederacy. The proclamation affected enslaved people in the Southern Confederate states but not those in the four slave states that remained in the Union. Further legislation was needed to abolish slavery in all states and U.S. territories, but this effort took time and diplomacy. The Thirteenth Amendment abolished slavery and involuntary servitude in all of the United States. It was ratified and adopted by a majority of the states on December 6, 1865.

USS *Hope* remained a blockader off Charleston, South Carolina, through mid-April 1864. In mid-June, Captain Rockwell was transferred to the sidewheel steamer USS *Yankee*. On July 7, John E. Rockwell married his fifth wife, Margaret C. Herold of Washington, D.C., and in August, USS *Kensington*, a screw steamer, returned from New York, where repairs were made. Captain Rockwell was ordered to leave USS *Yankee* and join the North Atlantic Blockading Squadron as commander of USS *Kensington* in mid-October. The steamer became a supply ship until December 1864 and was then designated a transport vessel from Boston to southern ports.

Captain John E. Rockwell in the Civil War, 1865. *Courtesy of Judy Rockwell Micoleau.*

On April 9, 1865, Confederate general Robert E. Lee surrendered his army of twenty-eight thousand troops to Union general Ulysses S. Grant at Appomattox Courthouse in Virginia. General Grant declared the war was over, and other Southern surrenders followed.

Captain Rockwell's eldest daughter, Olive Post Rockwell, wrote a letter to her uncle William Ayer, describing Hartford's reaction to the news of Confederate general Robert E. Lee's surrender.

Tuesday Evening, April 11, 1865

Dear Will,

Haven't we had glorious news this week! Suppose you voted for William A. Buckingham. Hartford went Union—something new for this place. The news of Lee's surrender reached here Sunday Evening about nine and at eleven everyone was up—bells being rung, bon fires, band's music out, and guns fired and all manner of things done. A very large procession was formed and passed through the principal streets, and in fact everyone was alive. Hope we will have a general jollification on Friday. Such is the talk now….

Write soon and excuse short letter. Burn this as it is only meant for one pair of eyes.

Yours in haste, Olive

In May 1865, the screw steamer *Kensington* was decommissioned, and five months later, Captain John E. Rockwell was honorably discharged on October 2, 1865. He returned to Essex with his new wife, Margaret. Olive Post Rockwell married James Woodbridge Hale in Essex on November 22, 1866, and the couple resided in Hartford for the rest of their lives.

From Sea to Land

In 1866, the Denison Shipyard in Deep River was still in operation near the town landing, and the yard produced a magnificent sailing vessel, schooner *William A. Vale*, in the yard beside Eli Denison's 1851 home. The 181-ton Schooner is depicted on the town seal and commemorates the town's shipbuilding industry. The Denisons built sixty-seven vessels in their yard from 1794 to 1882, wooden sailing craft until about 1870 and steamships, including paddleboat *Scorpio* built in 1865 and tugboat *Amos Clark*. The yard tried to stay afloat and incorporated new technologies in ship designs. It made the transition to steam-powered vessels, but business declined after the Connecticut Valley Railroad was established along the river in 1870, and railway ties cut through the shipyards. Fortunately, Deep River's economic base included ivory manufacturing and was not as affected as neighboring communities by the gradual decline of the shipbuilding industry.

Shipbuilding declined, but shipmasters continued to sail sturdy wooden packet ships. Lewis Post became second officer on the ship *Good Hope*, which sailed out of New York with a load of coal on a voyage to Hong Kong on Thursday, April 28, 1869, at 10:00 a.m. Five months at sea without access to news from home or letters from loved ones left Lewis struggling to maintain his emotional equilibrium. Lewis, then twenty-two years old, kept a journal of his voyage. He wrote pages about the weather conditions and wind directions, status of the sea, birds they encountered, his watches and a chief officer he was not fond of due to the mate's brutal manner of baiting

Denison Shipyard crew and schooner *William Vail* in progress, launched in 1866. *Courtesy of the Deep River Historical Society.*

and killing albatross. Lewis often reminisced about his days in Essex with his family and revealed his developing interest in Mary Rockwell, age nineteen, who at the time was living at her great-uncle Austin Lay's farm.

Lewis' developed a close relationship with Captain Daniel Kingsland Moore, master of *Good Hope*. They enjoyed lengthy conversations and endless battles of "chequers," backgammon, target practice and other competitions that took place nearly every day to pass the idle hours away. Lewis greatly admired Captain Moore's ability to write poetry and repeatedly wrote his diary in rhyming couplets to emulate Captain Moore's writing. On page 212, Lewis wrote:

> *I am every day confirmed in my opinion of Captain Moore, for I think he possesses the faculty of Rhyming to a great degree, for he writes off verses seemingly without the least effort. What a pity he had not commenced to write poetry years before. Think what the Human family has been deprived of for years gone by. Who knows but what his Genius might have soothed and quieted those Southern Politicians and Northern Abolishionists and*

thus prevented Our Bloody Civil War which has sent sorrow and desolation into the bosom of thousands of happy families.

I will take the liberty of abstracting a verse from his poem, that you may see that it is no overwrought sketch I am delineating.

"There was a Barque who tried with us to sail
But we kicked up our heels and showed him the tail
of the Ship Good Hope commanded by Moore
From the state of Lyme, Connecticut shore."

His intentions are I believe to have them published immediately upon arrival in Hong Kong, so look out for great excitement in the Literary World.

Lewis also had plans for his illustrated journal that concluded when they finally neared Hong Kong. On September 18, 1869, Lewis posted his journal to his mother, Meriah, in Essex, letting her know that he expected to be in Hong Kong for a few months. He was happy to add that the captain planned to return directly to New York, which they reached in 1870. *Good Hope* was a Black X ship that normally sailed to London. Its voyage to Hong Kong may have been experimental in order to establish a new trading destination. After the lengthy voyage to Hong Kong, Captain Moore left the ship, though not the Black X Line. He took command of the *American Eagle*, of which he was master in late 1867. Lewis Post was promoted to chief officer but decided to leave the sea in 1871 and make his living on land.

After Richard P. Williams retired from his work at New City Shipyard in Essex, he built a factory in 1867 at the site of his family's sawmill on the north side of the Falls River in Meadow Woods. Retired shipbuilder Nehemiah Hayden set up a new business there. He made coffin hardware in the Williams factory building for many years. In 1870, Richard Williams turned his home, the Williams Dam and his factory over to his sons. Richard Pratt Williams died on December 14, 1877, at the age of eighty-two. Nehemiah Hayden remained working at the Williams factory into the 1880s. He is listed on the 1881 map titled "A Bird's-Eye View of Essex, Connecticut," by O.H. Baily, as "No. 42. N. Hayden, Manufacturer of Undertaker's Hardware."

In 1870, Olive Rockwell Hale's half sister Mary moved to Hartford to live with Olive and her husband, James, along with James Hale's seventy-five-year-old mother, Julia Stiles Hale, who was blind and died the following year.

James and Olive Hale had one daughter, Edith Stiles Hale, in 1874. Both of James's brothers, Fred and Cornelius, survived the Civil War and continued their lives as private citizens.

The year 1870 was a good one for Captain John E. Rockwell. He was happily married, his three children thrived and his longtime friendship with Judge James Phelps was about to be honored. David Mack's shipyard was still in business after forty years on the banks of Middle Cove. John Rockwell was then fifty-five years old, had suffered the loss of one child and four wives, survived several harrowing commands, brought hundreds of immigrants from Liverpool and Antwerp to New York and commanded several vessels during the entire Civil War. John E. Rockwell was ready for retirement.

Captain Rockwell was determined to realize his dream of owning a schooner and sailing it in the Chesapeake Bay and surrounding rivers. He gave David Mack the commission. John Rockwell's schooner was one of Mack's last commissions and one of his greatest accomplishments. Built on the banks of Middle Cove near his home, schooner *James A. Phelps* was 112 tons, 88.5 feet long, with a beam of 27 feet, 3 inches, and a 7-foot, 1-inch draft. It was launched into the cove in October 1870, no doubt with great fanfare and celebration. Captain Rockwell was undoubtedly exhilarated while watching his beautiful new schooner slide down the ways into the cove from David Mack's slipway. As was customary, sixteen men owned shares in the schooner, including John E. Rockwell, Judge James Phelps, David Mack, Isaiah Pratt, Sam Comstock and three merchants from New York, but the New York Registry of Ships stated that John E. Rockwell was both owner and master of the schooner.

Schooner *James A. Phelps* was finished by the spring of 1871, and John and his wife, Margaret, were fortunate to enjoy a decade of sailing in and around the Chesapeake Bay. Captain Rockwell was familiar with all the ports and sailed the bay many times during his maritime career for purposes of commerce and trade and, later, for naval support and defense during the war. He knew those waters like the veins on the back of his hand. John Rockwell reveled in his retirement, doing what he loved most with his wife by his side, charting his own course each day in a bay he revered.

In late August 1871, the Rockwells received the tragic news that fellow Essex shipmaster Henry R. Hovey and others, including several Essex boys who served as Captain Hovey's officers, drowned in a storm off the coast of Florida. Captain Hovey, Amelia Champlin's half brother, was in command of the steamship *Ladona*, a steamship in the Mallory Line. *Ladonna* was lost

Launch of schooner *James A. Phelps* from Mack Shipyard, Essex, Connecticut. Builder, David Mack, 1870. *Courtesy of the Essex Historical Society.*

on August 20, 1871. First Mate Edgar C. Stevens survived the disaster and went on to command other steamships in the Mallory Line as well as the first oil tankers built for the Standard Oil Company.

On the advice of his mother, Meriah, Lewis Post proposed to Captain John Rockwell's younger daughter, Mary Ingham Rockwell. When Mary moved to Hartford to live with her sister, Olive, after the war ended, she became a dressmaker to help support the Hale household, which like many American families, suffered financial hardships and food shortages during the war.

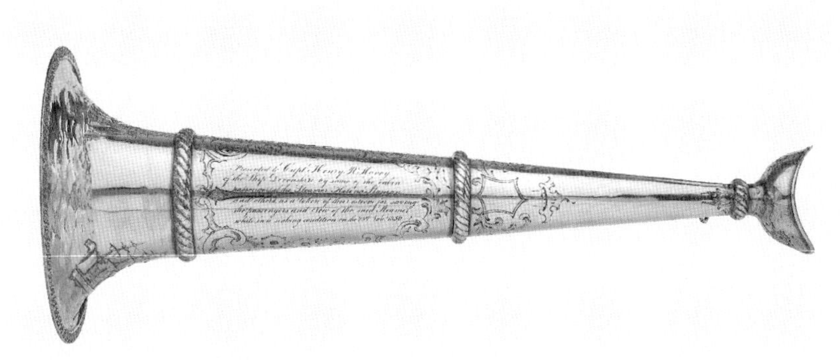

Captain Hovey's presentation silver trumpet, given to the captain by passengers of steamer *Helena Sloman*, for rescuing them and the crew. November 28, 1850. *Paul Foundation, Essex, Connecticut; photo by Cultural Preservation Technologies.*

Lewis Post was about twenty-six years old when he left the sea and decided to try his hand at manufacturing. After learning of the development possibilities in St. Louis, he decided to see what opportunities might await him out West. He went to St. Louis in 1872 and met his future business partner, a young developer named Lucian R. Blackmer who was familiar with the area. Lucien knew what land was available and where to find large deposits of clay. The two young men formed a partnership, and about 1878, they purchased a clay mine, now referred to as the Blackmer Post Mine, with the idea of developing a clay sewer and drainpipe business.

On December 10, 1879, Mary Ingham Rockwell, age twenty-nine, and Lewis Walter Post, age thirty, were married in St. John's Episcopal Church in Hartford, Connecticut. The couple left Essex and moved to St. Louis, where Lewis worked on developing Blackmer & Post Pipe Co. with his partner. The company began with a sixty-acre property that included the clay mine. The partners installed a furnace and constructed a factory that manufactured large clay sewer and drainpipes, eagerly sought after by city planners and developers. By the turn of the century, Blackmer & Post owned a clay mine, an office building, three factories and three large furnaces for firing the clay pipes. A rail system was used to move tons of raw clay to the processing plant and finished glazed pipes into the city.

On Lewis's office desk sat a thermometer encased in a brass ship's wheel. It was a prized possession that reminded him of his days at the helm. Being an officer in the Merchant Marine taught Lewis Post all the skills he needed to develop and thrive in his sewer pipe manufacturing company.

Lewis W. Post's office desk thermometer in a brass ship's wheel case. *Courtesy of the National Building Arts Center, St. Louis, Missouri.*

In March 1881, Captain John Rockwell and his wife, Margaret, sailed from Fair Haven, Connecticut, for the Rappahannock River in Virginia. The weather was unfavorable and cold, and John became ill with a cough and fever. Within days, he had pneumonia. His condition worsened, and when they reached the Chesapeake Bay, a doctor determined John was suffering from typhoid fever. John Everest Rockwell died at the age of sixty-six years on his schooner, *James A. Phelps*, in Chesapeake Bay, with his wife, Margaret, at his side. She applied for his pension for service in the Civil War and received payments until her death on November 9, 1904. Both John E. and Margaret C. Rockwell are buried at the National Cemetery in Arlington, Virginia.

In the spring of 1884, Lewis Post applied for a passport to Europe, and his partner Lucian R. Blackmer was his witness to the application. Among other information given, the document reveals that Lewis Post was five feet, five inches, with a mustache and medium complexion. His hair was brown and his eyes blue. He wrote that he was traveling abroad with his wife, Mary. The couple traveled by steamer and packed their clothes and other belongings in enormous steamer trunks in preparation for the voyage. At the time, Lewis's mother, Meriah, was living with them in Webster Grove, just outside St. Louis, and no doubt, she helped take care of their young sons, Malcolm Phelps, age four, and Walter Urquhart, age two. Meriah Post died on December 22, 1886, and her remains were brought back to Essex for burial beside her husband, David, in Riverview Cemetery.

Lewis and Mary Post's fourth child and only daughter, Marjorie, once remarked that her father learned to be successful in commerce and business from his years of experience as a ship's officer, especially trading a shipload of coal in Hong Kong for Chinese merchandise. When Marjorie was born in 1890, Lewis wrote to relatives back East, "She has the lungs of a bo'sun's mate!"

Captain William Urquhart was one of the last packet masters from Essex to leave the sea. His career was filled with harrowing tales and seemingly spiritually guided rescues at sea. He woke from a dream one night while in command of the ship *Tremountain* and told his officers to veer off course, as he suddenly knew there had been a collision. The evening was November 22, 1873. Miles away, the British ship *Loch Earn* collided in the darkness with the French mail steamship *Ville du Havre.* The *Ville du Havre* sank, and within minutes, 255 men, women and children were lost at sea.

Captain Urquhart's impression was correct. In the morning after the collision, he reached the horrific site of the tragic event and took the eighty-five surviving passengers and crew of the sunken steamship from the damaged *Loch Earn* and brought them aboard the *Tremountain*. The women were carried by way of a bo'sun's chair from the *Loch Earn* to the *Tremountain*. The legless chair was hauled forward by way of ropes suspended above the sea, like laundry taken in off a clothesline. Captain Urquhart gave the devastated survivors blankets, food and dresses that his wife, Emma, kept in a large trunk on the ship. Without hesitation, the captain brought the survivors to Cardiff, where they could make arrangements to be reunited with their families.

In 1879, while in command of the C.H. Marshall's Black Ball Line ship *Isaac Webb*, a 1,359-ton vessel of 185 feet, built in New York by William Webb in 1850, Captain Urquhart came across the British bark *Isolina* in a sinking condition. He rescued the crew and brought them to Liverpool. For his rescues and kind treatment of those whose lives he saved, Captain William Urquhart was given many valuable gifts of gratitude, including a gold chronometer and chain, a silver salver with gold trim and a sterling tea set and tray with engraved inscription.

Captain Urquhart's last command was on the *Isaac Webb*. While transporting a load of railroad iron from Antwerp to New York in November 1881, the cargo shifted during a storm, and the thirty-one-year-old ship foundered mid-ocean. Captain Urquhart put up a distress signal, and in time, a steamship rescued him and his crew and took them to Boston.

Captain William Urquhart retired from the sea in early 1882, following a merchant maritime career that spanned nearly three decades. He decided to operate an inn with his wife, Emma, and grown children: Kate Emma,

Ship *Isaac Webb*. Oil/canvas by John Hughes (British 1808–1880). Painted circa 1850. *Paul Foundation, Essex, Connecticut; photo by Cultural Preservation Technologies.*

an oil painter, and William Wallace Jr., an accountant, composer and piano player. His intention was to entertain his guests with music, paintings of ships and stories of his seafaring days. Captain Urquhart was then known as "an old salt" and advertised for travelers and boarders to make reservations for his inns in the late 1880s, including the Eagle Hotel in Essex and, later, the Melrose Hotel in New York.

William Urquhart's lifelong friend and brother-in-law Gilbert Williams's career as a merchant master ended later than his. After commanding packet ships through the mid-1880s, Captain Williams's final command lasted eight years on the steamship *State of Texas*. Following a stressful voyage in early February 1893, Gilbert Williams became ill and was brought to William and Emma Urquhart's home in Brooklyn Heights, New York, where his condition worsened. Captain Williams developed congestion of the brain, followed by spinal meningitis, and he died on February 23, 1893, at the age of fifty-eight. Gilbert Williams left his wife, Harriet, and two children, as well as his brother and sister-in-law William and Emma Urquhart, their two children and many relatives and friends in Essex, Connecticut.

Cousins William Urquhart and Lewis Post had a lifelong kinship, not only their blood ties but also an unbreakable bond from their years at sea together. In their retirement years, they maintained close ties, despite the miles that separated them, and kept in touch through visits and letters. The cousins met for the last time when Lewis Post, a retired pipe manufacturer living in Commerce, Missouri, visited William Urquhart at William's home office in Brooklyn, New York. William was writing a memoir on ocean travel and his days as a merchant mariner. Lewis was in his sixties and William in his seventies.

The walls of William's office were covered with oil paintings of ships that he and his father, John, commanded. One painting was of William's father's early packet ship *Lorena* in high seas. Another depicted *Tremountain* picking up survivors of the *Ville du Havre*. The scene shows a woman being transferred from the damaged ship, in a bo'sun's chair, across to the *Tremountain* and men in lifeboats beneath her searching the ocean for more survivors.

Gazing at one painting in particular, of the ship *American Eagle* under sail, Lewis and William reminisced about their first voyage together. Lewis thanked William repeatedly for not giving up the search for him, for going so far as to endanger the crew in order to save his life when so many mariners fell overboard and were never seen again. Lewis was reminded that William made a promise to his aunt Meriah, and come hell or high water, there was no way he would break it.

Dauntless No. 1

If there is one name that is most significantly associated with Essex boating in the late nineteenth and early twentieth centuries, that name would be *Dauntless*. Today, the Dauntless Shipyard stands at the head of Middle Cove. The fabled Dauntless Club enjoys its view of the river down on Novelty Lane. It is the current occupant of the legendary Hayden House, harking back to the town's earliest shipbuilders. There are three boats that have carried the name *Dauntless* on both the Atlantic and Pacific Oceans. Each of them has an important place in American maritime history and the legends of Essex.

The vessel that lent its name to other boats and a club that followed in its wake was, surprisingly, not built in Essex, although most folks think it was. The original *Dauntless* was designed by J.B. Van Deusen and built by the Forsyth and Morgan yard in Mystic, Connecticut. Originally rigged as a sloop in 1866, as the cloud of the Civil War was finally lifting from the country, it was initially named *L'Hirondelle* and owned briefly by S. Dexter Bradford. In 1867, it was lengthened to 121 feet and rigged as a schooner. It grossed 299 tons and had a 25-foot beam. The new owner was the larger-than-life newspaper czar James Gordon Bennett. He changed the name to *Dauntless*, a name that still resonates on the Connecticut River and beyond.

Bennett was to become the commodore of the New York Yacht Club and raced his beautiful schooner in British waters for several campaigns. His nickname in club circles was the "Mad Commodore," and he did his best to live up to the title. He was an avid sportsman; polo and ballooning were among his pastimes. He was also a dedicated tennis player. Bennett was notoriously bibulous and once caused the end of an engagement when he proceeded to relieve himself in the fireplace of in-laws-to-be during an elegant New Year's Eve soiree. The brother of the affronted and embarrassed fiancée gave him a sound thrashing with a horsewhip. The unrepentant Bennett then challenged his never-to-be-brother-in-law to a duel. Fortunately, they were both such lousy marksmen that they escaped unscathed from the confrontation—and wound up coming to an uneasy truce over a few beers after their near misses.

Prior to his acquisition of the *Dauntless*, Bennett was the winner of the first transatlantic yacht race. On a bet that grew out of a brandy-soaked evening at the Union Club, the young newspaper scion and two of his partners in revelry agreed to match their schooners against one another across the North Atlantic in December 1866. Although his boat, the *Henriette*, was the oldest and slowest of the three, the brilliant seamanship of Bennett's captain, an old packet skipper named Sam Samuels, gave him the win. It established Bennett as the world's premier yachtsman. He was feted by Queen Victoria and the cream of British nobility. The only downside to the race were the six crewmen who were swept overboard from one of the losing schooners. Such were the perils of a common seaman's life. The names of drowned men were lost to history, until a diligent researcher uncovered them in the twenty-first century.

Bennett swapped out *Henriette* for *Dauntless* and continued to follow his passion for long-distance yacht racing. In 1870, he raced *Dauntless* in a transatlantic match against Sir James Ashbury's schooner *Cambria* from Daunt Rock off Cork, Ireland, to Sandy Hook, New York. *Dauntless* unexpectedly lost that race, due to some questionable tactical decisions. But it didn't lose by much. That east–west ocean duel lasted twenty-three days, and Bennett's boat tailed the English yacht to the finish line by less than two hours. By then, *Dauntless* was the flagship of the New York Yacht Club Fleet. It would continue to race across the Atlantic and compete in both English and American waters. Bennett eventually sold *Dauntless* to John Waller, who owned it for only three years.

In 1882, Caldwell Hart Colt, the son of Samuel Colt, the inventor who created the revolving pistol, purchased the *Dauntless* from Waller. It was the young Colt who eventually brought the lovely schooner to Essex, where it found its true home. Known to his many friends as "Collie," Caldwell Colt is one of the most interesting figures of the nineteenth-century American yachting scene. His active, short life was (like many sons of successful

Gilded Age men) filled with languor, luxury, boats, drink, decadence and, ultimately, disaster. Collie Colt was heir to a vast fortune, a famous name and an industrial empire. But sadly, he also was an unwitting victim of what has come to be known as the "Curse of the Colts." His family, though it had aspects of great wealth, honor, accomplishment and glory, also had much more than its share of terrible tragedy.

This curse was transmitted to Caldwell though his father, Samuel. Sam's mother died of tuberculosis when the boy was only six. His father remarried and sired three half sisters for Sam and his brother. One of the girls died in infancy, one died as a teenager and the one who survived committed suicide. There was a lot of death and change for a sensitive, smart boy to handle. And young Samuel had his share of emotional ups and downs as he passed from childhood into his teenage years. He began to work in his father's textile mill, but it was decided that he should further his education.

Sam was admitted to Amherst Academy in Massachusetts. The prep school was selected so that he might study navigation, since from early in life he had a fascination with the sea. Unfortunately, he was tossed out of the school. The main reasons for his dismissal were inattention to his studies and poor grades. At his father's insistence, he then shipped out as an ordinary seaman aboard a ship named the *Corvo.* During his long watches on deck, his mechanically oriented mind became fascinated with the way the ship's wheel was geared. This fascination led him to put together the rudimentary ideas that ultimately evolved into the revolving mechanism that became the Colt .45.

The family's tragic vector cast a shadow over the life and fate of Sam's brother, John. John was a colorful character to say the least. He was a noted womanizer and ne'er-do-well. He was a notorious riverboat gambler, a fur trader, a soap salesman and a farmer. At one point, he was the debate coach for the University of Vermont. He finally found his niche as a traveling lecturer. John's favorite topic was double-entry bookkeeping. He wrote a textbook on the subject that was popular and profitable. Sam was hopeful that his wild brother had finally settled down, but such was not to be the case.

To Sam and his family's horror, John committed a grisly murder. The deranged Colt brother hacked a printer named Samuel Adams, who had printed his book, to death with a hatchet. The reason was a dispute over $1.35 discrepancy in the payment of a bill. John packed the corpse in a shipping crate that he ticketed for New Orleans from the port of New York. The rotting mess was found in the hold of the ship *Kalamazoo* before it left the dock. Detectives began to work backward to try to solve what could only be called a horrific crime.

John Colt was summarily charged and arrested. He was deemed sane enough to stand trial, although "insanity runs in my family" was a drumbeat during the trial. Throughout the court proceedings, Sam was in attendance almost every day and picked up his murderous sibling's legal fees. Sam saw to it that his brother lived as luxuriously as possible while he was imprisoned in the Tombs. Catered meals, cigars, plush carpet and a real bed were the order of the day in his cell. Needless to say, the New York and national press had a field day with such a high-profile case. Ironically, James Gordon Bennett Sr., father of the *Dauntless*'s future owner, led the charge to excoriate the murderer.

The Colt family black sheep was found guilty of first-degree murder. He immediately began the appeal process, but it was to no avail. His death sentence remained in place. But even that event, which should have been routine, was fraught with drama and intrigue. Attempts to smuggle him out of jail dressed as a woman failed. Hours before he was to hang, he married Caroline Henshaw, a pregnant Scottish woman. Rumor had it that the fetus she was carrying was really Sam's, not John's. After the wedding ceremony, a fire broke out in the Tombs. When the confusion settled down, John was found dead from a self-inflicted stab wound. It is believed that a family member slipped him a folding knife during the wedding. For years, conspiracy theorists surmised that John Colt did not die that day, but no proof of that was ever found. Suffice it to say, it was a terrible chapter in the annals of a family that had more than its share of tragedy.

Samuel Colt, the man who changed the social order of America with his Peacemakers, continued to have cascades of ill fortune heaped upon himself and his family. After some rocky starts, his enterprises became wildly successful. He reaped an extraordinary amount of honors and riches. But they were not to buy him happiness. He married Elizabeth Jarvis Hart in 1856. They had three children together, but only one, Caldwell, would live to reach adulthood. During the first year of the Civil War, which added to his estate immensely, Sam Colt died of gout-related complications. He left a pregnant widow and their three-year-old son, Caldwell Hart Colt. The baby Elizabeth was carrying was stillborn, and little Caldwell became the only-child center of her world.

Elizabeth was a doting mother, for sure, but she was also a shrewd businesswoman. She guided the Colt industrial empire with a firm hand and an agile mind. The Connecticut-based company continued to flourish and grow on her watch. In his twenties, Caldwell was named vice president of Colt Manufacturing. It was Elizabeth's fervent wish that he would lead the enterprise and fulfill the destinies and dreams of his father. The Yale-educated young Colt initially showed promise. At the tender age of twenty-one, he invented a double-barreled repeating rifle. Only sixty-five of the .45-70 Government–caliber weapons were made, which he distributed to friends. They are among the rarest and most valuable of Colt firearms today. But alas, to the concern and consternation of poor Elizabeth, Caldwell's path was not to be one of corporate ascendency. He preferred the profligate life of a scion of privilege, in the fashion of James Gordon Bennett Jr.

Yachting, fishing, hunting, strong drink, fine food and attractive women were the axis around which Caldwell's world spun. Central among these were boats. He owned five yachts in his lifetime. He spent millions (in today's dollars) on them. His devotion kept him on the water for an average of ten months each year. Of all the beautiful boats he owned, though, the *Dauntless* was the one closest to his heart. He took his mother across the Atlantic on it in 1882, and they proceeded to do their version of the Grand Tour. Mother and son scooped up European treasures and brought them back to Hartford to grace the expansive rooms of Armsmear, the grand family residence. Many of these artistic treasures may now be found in the Wadsworth Atheneum, Hartford's venerable world-class art museum.

Like Bennett, Caldwell also raced the *Dauntless* across the Western Ocean; in what has been called "an almost suicidal attempt to defeat the *Coronet*," Colt buckoed his crew like a clipper ship driver. He urged them on day and night through tempest after tempest. Owner Caldwell resisted the advice of his captain, Sam "Bully" Samuels, and sailed the sleek schooner into the teeth and eye of a hurricane. The boat was severely damaged. Crewmen had to man the pumps constantly to stay ahead of the leaks. Damage to the water tanks caused them to run out of fresh water. On a boat, in the middle of the ocean, that can be a circumstance that can often prove fatal. The resilient Commodore Caldwell responded to the dilemma by rationing champagne, ale and wine to keep his crew hydrated. The race was lost, but the crew wound up with a great sea story to tell around the yacht club bar. Colt was out $10,000, but it hardly mattered to him. He was doing what he loved. He was soon to become vice (no pun intended) commodore of the New York Yacht Club.

The schooner *Coronet* was owned by Rufus T. Bush, an industrialist and an oil refiner. His testimony against Standard Oil for its monopolistic railroad practices raised many eyebrows among his robber baron cronies. But Bush didn't care. He had a transatlantic race to win. His triumph was celebrated on the front page of the *New York Times*. After his victory over Colt, he sailed his splendid schooner around the world. The *Coronet* was then passed on to a series of owners, including the Kingdom, a secretive, cultish religious sect. The Kingdom owned it from 1905 to 1995 and sailed around the world on "prayer missions" and proselytizing voyages. One of these, to Palestine, ended in tragedy when six people aboard died of scurvy. The group's demented leader, Frank Sandford, was convicted of manslaughter as a result. Eventually, the International Yacht Restoration School acquired *Coronet* and commenced a complete rebuild and restoration. It survives to this day and can be seen docked on Thames Street in Newport, Rhode Island.

As he entered his thirties, Caldwell settled into a routine based on sailing the waters of Florida in the winter, cruising the Atlantic coast in the summer and bringing the *Dauntless* to Essex every autumn to take advantage of the excellent duck and rail hunting on the Connecticut River estuary. Every year between 1882 and 1894, the return of Colt and the *Dauntless* was an annual event that marked the changing of the seasons and the beginning of a serious social swirl in town. Caldwell would anchor his schooner in the deep water off Ely's Ferry Wharf. The playboy yachtsman would lavishly entertain his Hartford and New York friends over copious drinks and game dinners and raucous conviviality. Flares and lanterns were set in the rigging, and cannons and rifles would boom and bark. A makeshift band composed of his sailors played popular tunes, and the revelries would go on until dawn. As the days shortened and the chilly winds blew down the river, Colt would point his prized sailing yacht south and follow the birds to continue his pattern of hunting, fishing and pleasure-seeking in warmer waters.

It was there, off the coast of Punta Gorda, Florida, that the thirty-five-year-old bon vivant met his enigmatic end, fulfilling the prophecy of the "Curse of the Colts." Everyone at the time agreed that he died aboard his boat. The manner of his death, however, continues to be a matter of mystery, conjecture and controversy. Was it tonsillitis? Chicken pox? Cirrhosis? Did he fall overboard while drink taken? The most persistent answer to this riddle is that the unabashed rake was shot to death by a jealous husband. In many ways, this, even if it is not the case, is the most satisfying answer. It puts a period to the end of a life that is rife with irony. This would especially be the case if the cuckolded pistolero was using a Colt revolver. In any case, Caldwell Colt was dead at the age of thirty-five, leaving Elizabeth to mourn yet another untimely demise.

Caldwell Colt's schooner *Dauntless* decked over as a houseboat. *Courtesy of Essex Historical Society.*

After Caldwell's death, his mother turned the *Dauntless* into an Essex-based shrine to her son. She insisted that the schooner be kept "shipshape and Bristol fashion." Once a year, Elizabeth would come down from Hartford on the train and spend a week aboard the schooner to mourn her son and pray for his soul. She realized that as well made as the vessel was, its wooden construction rendered it somewhat ephemeral. To establish a more permanent edifice to memorialize her sailor son, Elizabeth erected a parish house at the Church of the Good Shepherd in Hartford. Earlier, she had the church constructed out of Portland Brownstone shipped up the Connecticut River on barges. It was built to perpetuate the memory of her husband and her deceased young children. She built it so her employees and their families would have a convenient place to worship and celebrate life's important transformations.

Elizabeth commissioned the parish house to be constructed adjacent to the church to honor her son's life and his love of the sea. The edifice still exists, and it is truly a remarkable monument to aptly celebrate Caldwell's yachting life. The exterior of the building has images of the *Dauntless* carved into its stonework, along with Poseidon, seashells, billowing waves and other maritime imagery. The interior was fashioned to reflect the below decks of Caldwell's yacht. Enshrined there are its ship's bell and other pieces of equipment. One has the feeling that they are below decks on the *Dauntless.* A large portrait of Collie dominates the main hall. He stands purposefully at the helm, eyes fixed on the horizon…his course set toward eternity.

Upon Elizabeth's death in 1902, the *Dauntless* was willed to one of her relatives on the condition that the schooner would never sail again. Permanently docked in Essex, it was leased to a New York sportsman to serve as a hunting lodge a for a few weeks a year. It was transformed into a houseboat and became a mainstay of the town's picturesque waterfront. In the winter of 1915, the valiant ocean campaigner, now sadly neglected, sank at its moorings into the cold, murky depths of the Connecticut River. The legendary Captain Thomas Scott, builder of the Race Rock Lighthouse, was tasked with raising it and towed the once-proud racer down Long Island Sound to his boat yard, where it was broken up, put on a train and shipped to Wisconsin for firewood. *Dauntless* no longer dominated the Essex waterfront, but its legend and name became an ongoing part of the elegant town's legend and lore.

The days of Colt's *Dauntless* in Essex were a time of transition in the construction of vessels on the Connecticut River. As the Age of Sail came to an end, many of the larger shipbuilding facilities quietly closed. A few steamers were built, but the Saybrook Bar at the mouth of the river limited their size and drafts. As shipbuilding declined, prime waterfront areas opened up, and local craftsmen realigned their focus from creating world-spanning ships to building pleasure and utilitarian boats. Boatbuilding could be a dicey economic proposition, but it also afforded individuals the opportunity to become their own bosses. A few of the shipyards metamorphosed into boatyards. The Mack yard, which had become W. Frank Harrison's boatyard, located at the head of the creek in North Cove, was purchased by Charles A. Goodwin, a prominent Hartford attorney, and renamed the Dauntless Shipyard, taking its name from Collie's fable schooner.

Dauntless Shipyard

The Dauntless Shipyard was an important nexus in the transition from the glorious era of the grand sailing ships to the yachting mecca into which Essex would morph. Under the direction of Ernest N. Way, the shipyard responded to the challenges created by that transition. A brochure noted that "in the 1890's there was a fierce competition among boat builders to design and build boats that were pleasing to the eye and fast sailers. As a result, some of the most graceful boats were built during that period. Aesthetics and skill were not the only important qualities: fast performance was a major factor as well."

In 1894, Way responded to that challenge. He built several Chic-class day sailers. Designed by Charles Mower, they were twelve-by-nine-foot cedar-on-oak beauties that featured a heart-shaped transom, teak seats, a bowsprit and an optional topsail.

Over the winter of 1897, the crew at Dauntless Shipyard stayed busy. They built fourteen Essex-class sloops. These boats were pencil-thin fifteen-foot racers. Under a cloud of sail, these ultra-sleek "'big, little ships' crowd more boat into their dimensions than has been achieved before. Good looks and fast sailing quality are among their bids for popularity," according to an advertisement at the time. They were designed and built by Ernie Way, who also made a splash in the day sailing community with his multipurpose design known as "Jolly Boats."

Per a period advertisement, the Jolly Boat was made "to meet the demand for a fast sailing small boat—one that could be easily raised and carried as a tender for larger boats—the DAUNTLESS Jolly Boat was designed by E.N. Way and built by the Dauntless Shipyard, Inc., Essex, Conn. The boat, itself, is splendidly designed, finely built, and from the skillful adaptation of the design to quantity production can me sold for a reasonable price. The Jolly Boat is a first-class tender, capable of carrying a very big load; it is light and easy to carry for two persons. It serves every purpose of a dinghy and is strong and able. With its Marconi rig raised it becomes a fast sailing little ship of powerful abilities, smart and a strong sporting acquisition… it sails very close to the wind and its towering mast reaches up to meet the higher air currents. In no sense is it a toy, or is it logy. Instead, it is smart and provides a world of sailing fun."

The name *Dauntless* was also appropriated by a group of Hartford sporting gentlemen. Most of them were from Hartford. Many maintained second homes near the mouth of the river in the Fenwick Borough of Old Saybrook, which was named for Colonel George Fenwick and his legendary wife, Lady Alice Fenwick. Fenwick was instrumental in the founding of Potopaug. In 1905, with Charles Goodwin as a primary motivator, these sports officially organized themselves as the Dauntless Club, no doubt to bask in the reflected glory of the departed Caldwell Colt and his schooner. Many of the original members transitioned over from the Hartford Yacht Club. HYC was the second oldest in the nation, preceded only by the New York Yacht Club, which always had a strong Essex connection, especially when Collie Colt was anchored there. Essex often hosted the NYYC's squadron as its splendid yachts made their annual summer cruise.

The Hartford Yacht Club built its Fenwick Station at Folly Point, near the river's mouth. It was an elegant two-story structure featuring a dining room, taproom, lounge and sleeping quarters for members. The club launch would greet members as they stepped off the Valley Railroad train from Hartford and whisk them to their clubhouse and boats. Unfortunately, this refuge of dedicated yachtsmen was consumed by fire. The building was a total loss. Only a couple of portraits and the brass cannon used to start and end races could be rescued. The loss of their clubhouse spurred several Hartford Yacht Club members to align themselves with the newly formed Dauntless Club.

Dauntless No. 2

The onset of World War I was met with a flurry of patriotic fervor among flag-waving Dauntless Club members. Several stalwarts paraded around the clubhouse packing .45 Colt pistols, in case a German sympathizer was foolish enough to try to invade their bastion of privilege. Charles A. Goodwin (a prominent Hartford banker and lawyer who ran for governor in 1910) and a group of fellow clubmen decided that the best way they could contribute to the war effort was to build a forty-foot (some say forty-five) dispatch boat. It was to be duly named the *Dauntless* and presented to the U.S. Navy to patrol for German U-boats in the waters of Long Island Sound. To this end, Goodwin commissioned Ernest N. Way to design the sleek motorboat. It was to be propelled by a four-hundred-horsepower engine to ensure that it could outrun any Germans it might encounter on the sound. Way's design later proved to be quite efficacious for the construction of rumrunners after the pall of Prohibition was cast over the country.

Described in local news as one of the biggest events in Essex history, the launching of the second *Dauntless* was a daylong celebration of the Stars and Stripes and all they stand for. Club members were joined by local and state politicians, naval officers, prominent citizens and other interested parties. Ladies shaded by parasols and men in dark suits and straw boaters gathered to mark the ritual transformation of the sleek craft from its perch on land to its proper element, the water. The town of Essex was aflutter in flags, flowers and bunting. Decorations from its Fourth of July festivities still gave the village a red, white and blue dressing from "one end of town to the other," mark this special display of national pride and support for the war effort.

The daylong festivities began when a parade stepped off in the Ivoryton section of town bound for the Dauntless Club at the end of Main Street. Several military marching units were joined by local musical aggregations. Governor Marcus H. Holcomb met the marchers at the club with warm greetings. He accompanied the procession up to Baptist Hill, where a flag-raising ceremony took place. The folds of Old Glory barely stirred in the sultry summer air, but it still caught the eyes of the onlookers as they squinted into the July morning sunshine. A few local dignitaries made some perfunctory remarks, and the august assemblage made its way back down the hill, where they gathered back at the club. The party then walked down Ferry Street over to the Dauntless Shipyard for the actual launching and its attendant rites.

The rapidly warming guests were then treated to an hour and a half of speeches as the sun rose higher over the town. The Honorable Charles Hopkins Clark of Hartford spoke on behalf of the builders and officially turned the *Dauntless* over to the Dauntless Club. The Honorable Walter H. Clark of Hartford then received the boat from its builders and thanked the builder on behalf of Roy T.H. Barnes, who, after all, had paid for it to be

The launch of the patrol boat *Dauntless*. This Ernest N. Way–designed speedster ferried officers from Boston to New York. *Courtesy of Essex Historical Society.*

constructed. The Honorable Morgan Gardner Buckeley then accepted the *Dauntless* on behalf of owner Barnes. Next up was the governor, looking very much like Teddy Roosevelt, except for the little beard under his lip rather than a brushy mustache. Many of the listeners' feet were probably hurting by then, but they still had had to endure the remarks on behalf of the navy by Rear Admiral William Sheffield Cowles. Cowles was the brother-in-law of the aforementioned Teddy Roosevelt, so the governor's resemblance to TR was sure to be noticed. Admiral Cowles, presciently, was one of the first naval officers to see the value of aircraft launched from ships as important advancements in water-based warfare. The reason for the protracted orations was the navy's new dispatch boat had to be launched at high tide, or there would not be enough water to float it down the creek out to the Connecticut River.

This timing's necessity was reflected in the message printed on the formal invitations the Dauntless Shipyard issued for the launch. The highly prized invites were headed with the aphorism, "Time And Tide Wait For No Man." This truism made the logistics of the launch time sensitive, lest the navy's newest addition wind up stuck in the mud. When the water had reached the level sufficient to provide buoyancy to the boat, the principal of Essex High School, William Leroy Burdick, read, in solemn, stentorian tones, Henry Wadsworth Longfellow's poem, "The Building of the Ship." The poem is a bit on the lengthy side, but when the principal got to the lines, "she starts—she moves—she seems to feel the thrill of life along her keel," workers were instructed to kick away the blocks and send the speedster down the greased ways to the water. Longfellow's trope was that of the water as a bridegroom and the ship as his virginal bride about to make her big splash.

All was set to go, the orator spouted the poem, the workers knocked away the blocks and the ensign was raised. Barbara Barnes, daughter of the project's funder, smashed a bottle of Moët & Chandon White Star champagne against the prow, a salute was fired from the cannon that once was on Colt's *Dauntless* and the band struck up, "Columbia, The Gem of the Ocean." But nothing happened! Captain Mack, who was holding the rope that tethered the boat to the land, in the midst of all the excitement, forget to let it go. Embarrassingly, the *Dauntless* remained firmly attached to terra firma. Sizing up the situation, club member James Lord Pratt roared out, "Mack! Let go of the damned rope!" His cheeks reddening, the chagrined captain realized his error and released the line; the blushing bride of a boat (according to Longfellow) finally slid into the anxious arms of her aqueous bridegroom.

Once the vessel was afloat, "The Star-Spangled Banner" was sung, and the invitees trooped back to the Dauntless Club for a reception and lunch honoring the dignitaries in attendance. Everyone enjoyed some light refreshments. The spanking-new speedboat made several rapid runs up and down the river to show off. A photograph taken from shore shows dark-suited gentlemen leaning forward, hats in hand, and two ladies sitting decorously on the cabin top holding on to their sunbonnets as they roared up the river. The yacht ensign flying from the stern and the Union Jack at the bow are both straight out. *Dauntless*'s rushing, white wake and the bone in its teeth suggest that it is going along at a pretty good clip.

The navy was not as taken with the *Dauntless* as club members might have expected or liked. According to one club member, "The boat was wonderfully successful, but she was ahead of her time and we couldn't get the brass hats to appreciate her." For whatever reason, the naval command did not think it would be suitable for the anti-submarine patrol work it was intended to do. Reluctantly, the officers accepted *Dauntless* as a gift. It was primarily used to ferry officers quickly between Boston, Newport, New London and New York. In most instances, it proved as fast or faster than the train service between these points. One passenger recalled a speedy trip between the Cornfield Lightship off Old Saybrook, Connecticut, and Fort Adams in Newport, Rhode Island, that took only two hours and twenty minutes.

Even though the navy might not have been overly enamored of the *Dauntless*, it added to the reputation of her designer Ernest N. Way. He was lauded as a visionary when it came to creating boats that could move quickly through the water. Ernie and his father owned a tobacco company that stored its broad-leafed product in a barn on Wethersfield Cove some thirty miles upriver from Essex. Exposure to the river as a lad led Ernie to a lifelong fascination with the design and function of boats. As he began to develop his own creations, he tested models in the cove. According to Brenda Milkofsky, doyen of all things Connecticut River–related, Way would pull his models along with a fishing rod to see how they behaved in the water. He added or subtracted pennies to change the weight and note the effects on the performance of his prototypes. Milkofsky notes, "He developed a theory of hull design that he adhered to for the rest of his career." Way told a local newspaper that "the logical under-water cross section of the hull should form the arc of a circle if the owner is seeking a combination of speed, strength, lightness, smoothness and safety."

Way was the superintendent of the Dauntless Shipyard and designed and built boats at that facility. He built sloops based on traditional Connecticut

River dragnet boats, which were used for shad fishing, originally, and then transitioned into pleasure sailing boats. These day sailers were beamy, shallow-draft boats, ideal for Connecticut River conditions. The *Nancy G*, which he built for Dauntless Club member Charles A. Goodwin (who was instrumental in creating the dispatch boat) is an excellent example of this design, with a sprit sail rig and bowsprit. But it was powerboats that gave Ernie his premier focus and lasting fame.

In 1906, Ernie designed and built *Columbia*, a fast, commuter-type boat, for fellow Hartford Yacht Club member Albert L. Pope. Pope was the scion of the Pope automobile and bicycle empire. At the end of the nineteenth century, his father had imported fifty bicycles from Europe and almost single-handedly initiated the bicycle craze in America. He built a factory to manufacture Columbia (hence the name of his son's boat) bicycles and later built some wonderful automobiles. But he was ahead of his time, and his cars were electric and could not compete with gasoline-powered engines. The lessons Ernie learned from building *Columbia* served him in good stead a few years later when he was called upon to create the dispatch boat *Dauntless* for the U.S. Navy.

Ernie Way found his true calling in creating small racing boats. Powerboat racing was in its infancy as he was coming into his own as a designer. Improvements in the development of internal combustion engines made them available for a growing variety of marine applications. One such was to propel small, light boats at rapid rates of speed. Way was so good at this that the rules of racing had to be changed in order to give competitors a chance at winning. Middletown, Connecticut, was selected to host a national speedboat race in 1930. Way had designed *Toy* to compete in and win the tournament. And win it did! It smashed a slew of world records, hitting up to an unheard of forty miles an hour on the straightaway parts of the course.

Toy broke more than twenty records in a two-day period. Its victories were so one sided that the rules committee reviewed the race and found that the lightweight speedster weighed only sixty-three pounds before the engine was attached. The stunned race commission decreed that henceforth boats must weigh at least 160 pounds to compete in that class. A saddened Ernie took his lightweight love back to Essex, and with the help of Frank Harrison, he replaced its planking with "heavier stuff" and returned to the racing circuit with his baby tipping the scales at a hefty 160 pounds. *Toy* was almost 100 pounds heavier than its previous manifestation, but remarkably, it still remained competitive. *Toy* can be found today housed in the small boat collection of Mystic Seaport.

The welcoming statement in the program of that race is an extraordinary document in terms of how the people who were creating a whole new type of boat racing viewed their role in history, especially the history of the Connecticut River. The racial slur it contains is unfortunate from our twenty-first-century lens, but it is indicative of the sensibilities of the citizens of Connecticut at that time. Its grandiose tone and high-flown language are entertaining to today's ear and the blatant sense of self-importance can only be marveled at.

> *On the banks of the lordly Quinnehtuhquet—"the long tidal river"—at the spot where stood the village of Mattabesett a full century or more before the birth of the Republic, we are gathered for the annual regatta of the outboard motor association. When the continent was yet a virgin waste, these waters were parted by birch bark canoes in the deft hands of the native redskins as they completed in pastimes that were prophetic forerunners of the tests of skill and engineering we shall witness during this regatta. Before our frontiers had been pushed to the Mississippi, these shores saw stout ships slide down the ways and venture forth to the far corners of the earth, laying the foundation for the commerce which built the industrial beehive of modern Connecticut.*
>
> *To our visitors who come from near and far to witness these events, to the contestants who will match their skill in this heartiest of sports, and to the delegates who represent in their Person the growth of interest in our streams as recreational centers, we extend cordial welcome.*
>
> *Middletown of today—an industrial community, the seat of a university, and the home of Men who rejoice in their heritage of river and ships—greets you and bids you enjoy what they have been honored to provide for you. The Connecticut River Regatta Association,*
> *E. Kent Hubbard, President.*

Under the leadership of Ernie Way, Dauntless Shipyard established itself as a leading builder of both pleasure and utilitarian craft. The business letterhead in the 1920s announced that they were "Builders of Yachts, High Speed Motor Boats and High Class Work Boats. Any vessel in wood up to 100'." This pronouncement was to be borne out by a succession of splendid sea craft created at the yard during the early decades of the twentieth century. Some of their boats are still sailing almost a century later. Today, Dauntless Shipyard, while no longer in the boat-building business, remains an excellent repair, sales and storage facility from its proud perch at the head of North Cove.

Major Smyth and Dauntless No. 3

In 1925, the redoubtable Major William Smyth took charge of the operation there and proceeded to build some of the finest sailing yachts ever to grace the waters of the Connecticut River and beyond. The major could be seen and heard roaring about the town in his Dort roadster at all hours of the day and night. The affable mariner was not only a superior boat builder but also a shrewd administrator who was able to develop a business model that began in the heady days of the Roaring Twenties. His systems were able to sustain themselves during and after the Great Depression and continue into the present day.

Although his tenure in Essex only lasted six or so years, Smyth was able to leave a lasting imprint on the town and, in that period, create some memorable sailboats. He is such a unique and personable part of American maritime history that his story deserves retelling. William Smyth was born into the family of a prominent physician in Dublin, Ireland, at the end of the nineteenth century when the world was in transition from the Victorian to the Modern Age. From these privileged beginnings, he went on to live a life devoted to the sea and the boats that sail on it. Connecticut is fortunate that this extraordinary sailor-man came to call the Nutmeg State his homeport for most of his life.

Smyth's seagoing career began when he attempted to establish an offshore fishing fleet for Ireland. Since time immemorial, fishermen can be quite territorial in terms of their prized spots. This holds true for nations as much as for individuals. In Smyth's time, Ireland, Scotland and England were engaged in some heated conversations and confrontations as to who could fish where and when and how much. When these differences of opinion escalated into violence, Smyth began to rethink his commitment to the fishing industry. A bullet whistling past his nose in the wheelhouse of a trawler put a period to his fishing career. He never knew the identity of whoever shot at him, but he decided to give fishing a bit of a rest.

In a classic example of out of the frying pan into the fire, though, the adventurous young William enlisted in the Irish Horseguard as World War I was heating up. His regiment started out with ten thousand soldiers and finished the war with only three hundred, so it was a grim business in which he found himself. While in the service, Smyth made a name for himself as a welterweight boxer. He was wounded in combat twice, and eventually he was promoted to the rank of major. Major became the title by which he would be known throughout his long and interesting life. When the horrific hostilities finally came to a halt, the newly mustered-out officer cast about to see where the next chapter of his life course would take him.

The answer was back to sea. The major crossed the Atlantic and took up seafaring as the captain of schooners for the Hudson's Bay Company. His career there was almost ended by a psychotic ship's cook wielding a meat cleaver, but he was able to disarm the madman and live to sail another day. Smyth's schoonering experiences instilled in him a lifelong love of wooden boats. He was able to follow the trajectory of that love into a career that epitomized him as one of the most interesting and important characters in yachting history. He turned his passion into his career.

The major came to Essex by way of the Minneford Yacht Yard on City Island, New York. He later went on to finish his work in Connecticut waters eastward down Long Island Sound in Mystic. We shall discuss in detail some of the splendid vessels he constructed at the Dauntless Shipyard, but first let's look at some of the highlights that made him so invaluable to boating history. Outstanding among these was his love of historically important ships and boats. Smyth was instrumental in the maintenance and preservation of two iconic American sailing craft.

One was the last wooden whaling ship, the *Charles W. Morgan*. Smyth had a longtime relationship with the Howland family of Massachusetts. The Howlands made their fortune in the whaling and shipping trades, and the *Morgan* was one of their ships. After its working days were finished, it wound up being anchored off Round Hill, the estate of Edward Robinson Howland Green. Better known as Colonel Ned Green, he was the son of Hetty Green, the Witch of Wall Street. Hetty is still listed in the Guinness Book of World

Records as the Stingiest Person in the World. She left Ned with more money than he could possibly spend, but he sure gave it the old college try. Ned brought the *Morgan* to Marblehead to accentuate his view. But it fell into disrepair. It was Major Smyth who reconditioned the vessel and fitted it out so that it could be towed from its berth in the mud to Mystic Seaport. The old whaler recently underwent a total restoration and is seaworthy once again and can be seen at the Seaport.

Major Smyth was associated with another famous boat. At one point, he was the owner of the *Western Republic*, the boat in which Howard Blackburn made one of his amazing transatlantic crossings. Blackburn was a larger-than-life seagoing legend. He started out as a Grand Banks doryman. Separated from his schooner in a blizzard, he rowed for three days until he reached the shores of Nova Scotia. By this time, his hands had frozen to the oars, his feet were badly frostbitten and his dorymate had frozen to death. As a result, Blackburn lost all his fingers and toes, but at least he was alive.

The loss of his digits didn't slow the redoubtable sailor down much, though. He became a successful saloon keeper and businessman in Gloucester, Massachusetts. But with salt water flowing in his veins, the vigorous bloke grew restless and decided to return to the sea. Without fingers and toes, he completed two single-handed crossings of the Atlantic under sail. He also took a dory up the Hudson, across the Erie Canal, down the Mississippi and up the Atlantic Seaboard back to Massachusetts. Major Smyth did history a service by preserving one on his boats. The *Western Republic* proved to be a bit crank as a sailer, and Smyth was found it to be quite a daunting challenge to teach a good friend's son to sail aboard it.

While he was the driving force behind the Dauntless Shipyard, the major was responsible for building some elegant, seaworthy and beautiful yachts. He built the first design ever created by Winthrop L. "Wink" Warner, one of the Connecticut River's very best marine architects. This boat was christened *Felisi*, and it was built for Commodore T. McDonough Russell of the Middletown Yacht Club. It was a fifty-two-foot ketch and launched on September 30, 1929, just a month prior to the Black Friday crash of the stock market that sent the world cascading into the Great Depression. *Felisi*'s function as a pleasure yacht was interrupted when it served as a U.S Navy patrol boat during World War II, on the watch for German U-boats. It stayed afloat through the 1980s and gave its seven owners thrills and adventures galore.

Under Major Smyth's watch, a complete class of one design, racing daysailers, was created. They were named, appropriately, Dauntless 21s.

The fast sloops were both designed and built at Dauntless Shipyard. The trim racers were twenty-one feet in length overall, with a four-and-a-half-foot beam. They drew three and a half feet. With a keel they weighed eight hundred pounds and checked in at three hundred without. They made their debut splash on the racing scene at the 1931 Motorboat Show (of all places.) The demo model that was exhibited there was subsequently shipped to Bermuda. They became popular on that island and were raced there for many years. Dauntless also built a series of Madison Beach designs of twenty-one and a half feet in length. Designed by Sparkman and Stephens, they were primarily used on the sheltered waters of Long Island Sound.

Although his tenure in Essex was rather brief, from 1925 through 1931, Major Smyth was able to fashion and construct some spectacular sailboats. For the most part, he concentrated on building John Alden designs, along with the occasional Sparkman and Stephens. While he was in Essex, the Dauntless Shipyard built at least six schooners, four sloops, a ketch, a cutter and a yawl. His output during this time also included a thirty-foot motorized day cruiser and a fast fifty-five-foot motor yacht named *Viola*. Listed as Alden design no. 487, it is described as a "Special Boat" in the Alden index. *Viola* was launched in 1930. It had very sharp lines and was powered by a gigantic Sterling Viking 8-cylinder engine that developed a hefty 565 horsepower. Speculation at the time suggested that it was designed, built and used to surreptitiously run rum up the Connecticut River from the schooners anchored on Rum Row outside the law's purview off the coast of Long Island.

He occasionally built designs by other marine architects, but Major Smyth's most popular yachts during his tenure at the Dauntless Shipyard were classic John G. Alden designs. Guests were invited to the launch of one these named the *Golden Hind* on June 12, 1926. Their invitation was headed with the words, "FOR ME THIS LAND, THAT SEA, THESE AIRS, THOSE FOLKS AND FIELDS SUFFICE." The source of this quote is lost to history, but the invitation noted that the schooner would leave the land and take to the water at two o'clock in the afternoon "Deus Volens." Just as the invitation to the launch of the dispatch boat *Dauntless* had intoned almost a decade before, "TIME AND TIDE WAIT FOR NO MAN." Apparently time and tide cooperated and the *Golden Hind* slid smoothly down the ways at the appointed hour.

The *Golden Hind* was named after Sir Francis Drake's ship that circumnavigated the world on a privateering voyage in the sixteenth century. The twentieth-century version was built at the request of the

Launch of the *Golden Hind* at Dauntless Shipyard. After an adventurous career on many of the world's oceans, it was wrecked on Maui. *Courtesy of Essex Historical Society.*

aforementioned Charles A. Goodwin, a dominant figure in early twentieth-century Essex yachting circles. In a nod to history, the owner inscribed all the boat's glassware and china with Drake's family crest. It was Goodwin who was instrumental in the development of the dispatch boat that the Dauntless Club gifted to the navy. He was renowned among his peers for his seamanship skills. In fact, his lovely new schooner was not equipped with an auxiliary engine. This allowed for a lot of extra storage space, but it also bespoke a great deal of confidence on the parts of both the builder and the owner that it would sail well under all conditions. Goodwin's fellow club members always enjoyed sitting on the clubhouse verandah to watch him pick up his mooring on Sunday evenings after the weekend's sailing.

An engine was added in 1935, when it was sold to Franklin M. Haines, a New York real estate executive and commodore of the Fishers Island Yacht Club. Haines renamed it *Marita*, and it became one of the three Alden schooners he owned in his lifetime. Haines sailed *Marita* to a fleet first in the 1936 New London–Marblehead Race. He then had Phillip Rhodes alter its rig. The foremast was heightened, the bowsprit shortened and the main boom raised and cropped. These changes allowed *Marita* to carry a larger genoa, without a change to its racing rating to make it more competitive against newer boats.

The schooner's historic name, *Golden Hind*, was returned to it in 1941. In 1958, it competed in the Bermuda Race. The aging beauty was then sold to a new owner on the West Coast and spent several years cruising in the South Pacific. *Golden Hind* had matured a bit, of course, but all who had the pleasure to see it or sail aboard it, agreed that it was a beautiful boat. The schooner was recognized as a fine monument to the Dauntless Shipyard's craftsmanship and Alden's design skills. In 1974, *Golden Hind* participated in an informal regatta with two other Alden schooners off Papeete, Tahiti, racing one last time alongside *Mayan* and *Mariah.* Onlookers remarked on what unforgettable sights they were with "bones in their teeth and rails awash." *Golden Hind* continued to cruise the South Pacific well into the end of the twentieth century.

In 1931, Major Smyth constructed *Cock Robin*, a large sloop, for John P. Elton of Waterbury, Connecticut. Although Smyth said it was designed to cruise with a crew of only two, the huge mainsail must have been quite an imposing chore to handle when the breeze was up. *Cock Robin* was fifty-six feet at the waterline with a twelve-and-a-half-foot beam. One ongoing complaint about the boat's rigging was the lack of space between the headstay and the forestay. This made it quite difficult to come about when one of the larger jibs was set. But on a beam reach, it could put her shoulder into it and roar along at an impressive clip.

Cock Robin had a ton of lead ballast in its keel that allowed it to carry a great deal of sail area. But even with quite a lot of ballast, the sloop was difficult to handle comfortably. Owner Elton requested that Alden look into the possibility of modifications that would make it easier to handle without sacrificing too much speed. The solution was to convert it to a yawl rig. After Dauntless Shipyard carried out these modifications, *Cock Robin* then carried 825 square feet of canvas in its main sail, 156 square feet in the mizzen and 276 square feet in the staysail. When the crew set its Yankee, another 357 square feet was tacked on.

Owner Elton made a great personnel decision when he hired one Captain Nielsen to be his professional sailing master. Nielsen lived aboard *Cock Robin* for over twenty years. Among the most exciting hours of those two decades must have occurred in 1938. The skipper stayed at his post on the splendid vessel as one of the worst hurricanes in New England's history bore down on the coast of Connecticut. Nielsen orchestrated a whirlwind of activity as he belayed, battened, shifted anchors and fended off other boats. Essex

Harbor was transformed into a maelstrom of chaotic waves and wind. His Herculean efforts were to prove exemplary. After the tempest finally blew itself out, *Cock Robin* was the only boat in the waters surrounding Essex that did not sustain any damage.

Chester Bowles, one of the preeminent citizens of Essex, purchased *Cock Robin* in 1947. He renamed it *Mara*. Bowles was in innovator in the advertising field. He and his partner, Batten, grew their small agency into a Madison Avenue powerhouse that generated enough financial resources to allow Bowles to indulge in his yachting habit. He was the creator of the radio soap opera, and that medium gave his agency the opportunity to develop specific, target audiences, geared toward identifying the people most likely to purchase the products his clients wished to market. His dream was to take his rechristened schooner on an extended cruise to the West Indies and beyond.

But Bowles was encouraged by his many and varied admirers to apply his prodigious intellect and interpersonal skills to public service. He put his cruising dreams on hold and transitioned into government work. He was elected to be the governor of Connecticut in 1948 and from that office moved onto the national and international stage. He served as undersecretary of state and then served two stints as U.S. ambassador to India and Nepal. He played several key roles in shaping America's political and economic positions in the post–World War II international arena. He put his dream on hold.

His political obligations did not afford him the luxury of extended cruising due to time constraints. But he slipped away for a sail whenever the opportunity presented itself. History does record that he made it at least as far as Eggemoggin Reach in Maine. It was there, way Down East, that his boat came in contact with the Deer Isle Bridge. This meeting of wood and metal clipped eight feet off the top of *Mara*'s mainmast.

Mara was nicely appointed above and belowdecks. It was originally fitted out with a thirty-five-horsepower auxiliary engine that was placed up forward of the galley. Bowles had it replaced with a Gray twenty-horsepower gasoline engine. He located the new power plant aft to make it accessible from the cockpit, which was a plus. However, this forced him to do away with the after cabin, which left it a bit short of accommodations for a boat of its size. The aft stateroom was restored in 1965 when Harold Dennis purchased the boat and renamed it *Cock Robin.* The boat was a bit beaten up by this time and underwent an extensive rebuild. It could still hold her own on the racing circuit against much younger boats, however.

In the last half of the 1970s, *Cock Robin* was given a new life as a commercial swordfishing boat. It was fitted out with a long bow pulpit that enabled a harpooner to strike his or her prey from an advantageous angle. In 1980, it underwent another rebuild. Some floor timbers and ribs were replaced, and *Robin* was given aluminum masts, much lighter than its traditional wooden sticks. This reduction in weight aloft made it possible to remove 1,800 pounds of lead ballast from the keel. It switched from fishing back to racing and won the King's Cup Race from Santa Barbara to Los Angeles.

In 1981, the name was changed again. This time, the boat became known as *Scottish Fantasy II.* It made a fast fourteen-day passage from California to Hawaii and spent time cruising the islands. At one point, it made eleven and a half knots roaring along in a fifty-knot gale under its Yankee, staysail and mizzen. Crew members speak of that experience as one of the most fabulous sailing moments of their lives. Time finally caught up with the Smyth sloop, but not before it enjoyed many decades of cruising and racing experiences that created lifetime memories for all who had the pleasure of sailing it.

Another beauty that was put together by Major Smyth in Essex was *Teragram*, Alden design no. 397. It was built for George W. Mixter, a man who dedicated much of his life to all things nautical. He was the author of *The Primer of Navigation*, among other publications. According to Google Books, his *Primer*, "Since 1940 has been the most relied upon guide for sailors and boaters throughout the world. Within three years of its original publication, the book went through 15 printings. It has gone through 15 reprints since then." The cover photograph of the first edition shows Mixter looking suitably nautical standing on the deck of *Teragram.*

The fifty-five-foot schooner's name was that of Mixter's wife, Margaret, spelled backward. *Teragram* campaigned in several Bermuda races. It finished second in its class in 1930 and fourth in 1932. Like many Dauntless-built Alden schooners, it wound up in the Pacific. *Teragram* had many races with its sister schooners off the coast of California and in the Hawaiian Islands. Unfortunately, it met an ignominious end on the beach at Kona, where, according to the captain at the time, "her owner turned her in to matchsticks…it was an insurance job." One of *Teragram*'s crew was so distraught at the supposedly intentional wrecking that he gave the owner a severe beating. It seems a sad coda to the life of such a beautiful boat. While it was in its heyday, though, it often raced against the schooner that carried the proudest of all the Essex names.

That name, of course is *Dauntless.* Like Colt's schooner and the navy's dispatch boat, one of Major Smyth's excellent Aldens bears that historic nom de mer. It was built for Horace Merwin in 1930. At the time, Merwin was, like James Gordon Bennett before him, commodore of the New York Yacht Club. Smyth certainly built this *Dauntless* to last. It is still sailing today, almost ninety years since its hull hit the Connecticut River. Merwin idolized both Bennett and Collie Colt. They provided the inspiration for him to commission Dauntless Shipyard to build him a boat that could hold its own on the racing circuit. And hold its own this last iteration of *Dauntless* certainly did. It was fitted out with thirty custom-made Ratsey Lapthorn sails, and it carried more sail area than was usual for Alden-built boats of that era. It did have classic Alden characteristics, however, such as a traditionally graceful sheer, moderate overhang, a short spoon bow and an oval transom. However, it was built with less beam and more freeboard than many of its predecessors. With a fore-gaff rig, its foremast was a bit lower than was usual, also.

Commodore Merwin wasted no time putting *Dauntless* into competition. It won several North Atlantic and Bermuda races. When it finally made its way to the Pacific coast, where it remains today, *Dauntless* won the San Francisco's Master Mariner's Regatta. Like its sister schooners, *Dauntless* went through several owners and name changes in its life. From 1937 to 1960, she was known, exotically, as *Crishaza.* For a brief time (1960–61), it was fittingly named *Grace* and owned by Louise Grace, owner of the Grace Shipping Lines. It was during this period that the boat played a significant role in oceanographic research. It served as a platform to test a newly invented telemetric instrument that revolutionized undersea seismology.

Dauntless's original name was returned to it in 1962, when it was donated to Columbia University. It was berthed for a time on the Hudson River. From there, it took a turn down in Brazil, where it was rumored the craft's owner was a Spanish royal exiled from his Iberian home. It wound up back in the East Coast and fell into disrepair, settling unhappily into a mudbank. But a man named Bob Sloan rescued it. Sloan was famous in sailing circles for designing and building the schooner *Spike Africa*, one of the last "working" tall ships, as its current owners describe it. Sloan singlehanded *Dauntless* through the Panama Canal and brought it to Los Angeles for a total restoration. He chartered it between Newport Beach and Catalina Island for a few years without an engine.

The proud schooner was purchased in 1984 by Paul Plotts, a California restaurant owner who continued to keep it up and improve it. He raced the boat competitively in both California and Hawaii. Potts said, "I have never had a boat like *Dauntless,* she's large, a traditional old beauty, yet handles

John Alden–designed schooner *Dauntless*. Built by Major Smyth at Dauntless Shipyard in 1930, this beauty is still sailing. *Courtesy of Doug Domainie.*

like a baby. I can sail comfortably and confidently on her in just about any weather conditions." It measures sixty-one feet on deck and forty-three feet at the waterline. Just as Smyth built it, it is double-planked mahogany over oak frames. The masts are varnished spruce, and decks are teak. It often hits thirteen knots if the breeze is hearty.

Dauntless's stout heritage came in handy in 1991, when it was caught in the Moloakai Channel in twenty-five-foot seas. The hard, square seas of the channel took a toll on its stem and forward planking. The old girl sustained an injury that almost proved fatal past the point of no return on its way back to California. At one point, the boat was taking on 1,500 gallons of water an hour. The crew tore up the cabins and bunks in attempts to stop the leaks; both pumps were pushed past their limits. *Dauntless* was clearly sinking. Only the deus ex machina–like appearance of a Coast Guard C-130 airplane that dropped pumps and supplies to the stricken boat prevented a watery end to its grand career. *Dauntless* lived to sail another day.

It was recently listed for sale. The brochure touted a major rebuild, enumerated many new pieces of equipment and alluded to its luxuries. Let's hope that it will continue to grace the world's waters for a good long time to come. This sentiment is well reflected in a poem that a crew member wrote:

> *John G. Alden designed this yacht, / O Lord, she's mighty purty. / She was the pride of the Dauntless Yard. / in the year of Nineteen Thirty. / A classic staysail schooner, / she's wooden to the heart. / Mahogany Oak with strong teak decks, / she'll never break apart. / With varnished rails and snow white sails, / rigging strong and sturdy, / she still sails proud across the seas, / thru weather fine and dirty. / So give three cheers to the* Dauntless, */ a loud hip hip hooray. We'll always love ya,* Dauntless, */ you're still a queen today.*

Major William Smyth left the Dauntless Shipyard in 1931. He moved eastward down Long Island Sound and established himself as the prime mover at the Mystic Shipyard, where he enjoyed the rest of a fulfilling career working with some world-renowned yachts on the Mystic River and Fisher's Island Sound. In the grand tradition of mariners across the globe, he was fond of a glass of rum now and then. He called them "heave aheads" and downed them with alacrity. He was a fixture in the Mystic community and would carry on intermittent conversations with friends, picking up threads and themes immediately even though the last time they spoke was weeks prior.

The Major, unfortunately, broke his hip in his mid-eighties. He did it with some panache, however. He was scrambling aboard his twenty-eight-foot motorsailer *Quest* when he missed a step and fell. He was carrying a case of beer at the time. He wound up in the hospital and was given anesthesia. As he was emerging from the fog, he began to relate to the attending physician an analogy about his ether experience. He compared it to the wonderment of water ousels, the birds who can fly through waterfalls. The doc was not a literary man and thought the old salt had dementia. The Major dismissed him and proceeded to re-rig the traction apparatus holding his leg in the air.

After Smyth's departure, Dauntless Shipyard began to wind down its boatbuilding component. It concentrated increasingly on repair services, maintenance, storage and brokerage. Today it remains a fixture of the Essex community. It pronounces itself to be "recognized for personal and professional service, the Dauntless crew is skilled in the maintenance, service, and restoration of recreational and competitive yachts. Uniquely, this yard provides clients with customized plans designed to help each yacht owner develop a plan that is specifically tailored for them." The days of Ernie Way and the major may be long gone, but the name *Dauntless* is still proudly spoken in the village of Essex.

Essex Boat Works

While the Dauntless Shipyard was building pleasure yachts in the twentieth century, its neighbor just down the creek, the Essex Boat Works, concentrated on the construction of working boats. Essex Boat Works built a tugboat, a Stonington dragger and at least one lobster boat that was yacht-like in its attention to detail and appointments. The yard had a marine railway that allowed it to haul large vessels (the Hadlyme-Chester Ferry and Mystic Seaport's schooner *Brilliant*, for example). Like the Dauntless Shipyard, Essex Boat Works eventually gave up building boats to concentrate on renovation, storage and repair. But the steel lobster boat they put together is still hauling in the tasty crustaceans at Point Judith, Rhode Island, and the tugboat it built was a mainstay on the Connecticut River for decades.

The wood tugboat crafted at Essex Boat Works was named the *Chaco* for reasons that are now lost to time. It was built just prior to World War II for John G. Holbrook of Westbrook. At forty-nine feet, it was just a smidge too small to be commandeered into naval service. According to one of its subsequent owners, *Chaco* was the "last wooden vessel of any size built in Essex." It was powered by a D17 Caterpillar engine that propelled it up and down the Connecticut River for years as the crewmen built docks, drove pilings, retrieved sunken vessels, moved heavy equipment and, essentially, did whatever work came their way so they could make a living on the river they loved.

During *Chaco*'s heyday, there was heavy commercial traffic on the Connecticut River. Tow boats pushed and pulled barges deeply laden with petroleum products, and big tankers plied the river as far upstream as Hartford. These large vessels required sturdy docks and dolphins to keep them secure to the riverbank as they discharged their flammable cargos. New construction and repair of existing docks kept the *Chaco* and its men busy summer after summer. It would routinely steam eight or nine hours up to Hartford in order to build and maintain the oil docks of the Hartford Electric Light Company. Ned Libby, one of *Chaco*'s crewmen, who eventually became its owner, would have to get up very early in the morning tied up there, in order to fire up the steam engine that powered the crane on the barge.

It was at Hartford Electric Light that *Chaco* suffered one of its two submersions. A poorly secured coal barge broke loose from its moorings, careened down the river and smashed directly into the tug, which was tied to the dock, unaware that danger was bearing down upon it. Fortunately, no one was seriously injured, but *Chaco* had some planking stove in and sank unceremoniously at the dock. Its other sinking occurred one chilly December. The doughty tug was summoned upstream from Essex to do some repair work to the dolphins on the Hadlyme side of the Chester ferry. The crew knew that *Chaco* had a sprung plank, but since it was well above the waterline, they did not concern themselves with it too much, and up the river they went. They had not taken into account the weather forecast. While the tug was tied up overnight in Hadlyme, two feet of heavy wet snow blanketed the region. The weight of the dense snow pressed the hull of the tugboat down just enough to allow the river to find its way into the open seam. When Ned and the rest of the crew showed up for work the following morning, their office was at the bottom of the river.

Despite the dunkings and the hard knocks delivered by its chosen means of making a living, *Chaco* always kept up a well-maintained and polished appearance. It maintained a close relationship with its makers at Essex Boat Works, doing in-kind jobs for the yard. In turn, Stu Ingersoll, the owner of the boat works saw to it that, in Ned's words, "He kept the boat spiffed up." In 1990, Steve Cryan painted a picture of the *Chaco* that graces the north wall of Ned's living room today; it shows the hardworking tug in a cradle out of the water in front of the Essex Boat Works shed. A light mantle of snow covers the wheelhouse, cabins, rail and deck. The paint and brightwork look sharp. There is a large H painted on its stack in honor of John G. Holbrook, the original owner for whom the tug was built. It also sports a small sign advertising "Connecticut River Dock and Dredge," the company under whose aegis it operated for years.

Toward the end of *Chaco*'s time in Connecticut, Ned Libby became its owner. The tug was showing its age a bit by then, and he related that he had to "keep a bucket of sawdust handy, to slow up the leaks." But the old girl still looked good with its pilot house brightly varnished and shining. Ned reluctantly decided to put it up for sale. He listed *Chaco* with a local yacht broker and went about his business. He was in no hurry to sell it, but if the opportunity presented itself, he probably would. Ned was somewhat surprised when his broker informed him that there was a gentleman from New Jersey who was interested in purchasing the tug.

Ned got in touch with him and began to suspect that something was a bit fishy when the prospective buyer declined to have the boat surveyed before he bought it. This was quite unheard of in the boating world. Marine surveyors perform the vital function of detailing all that is good and bad in a boat so that the buyer knows exactly what she or he is getting. In order to protect himself from possible litigation in the future, Ned had the buyer supply a letter from an attorney stating that possible new owner declined the opportunity to have the *Chaco* surveyed before he took it home to New Jersey.

Tugboat *Chaco*, in a 1990 painting by Steve Cryan. It worked up and down the Connecticut River for decades. *Courtesy of Carol and Ned Libby.*

By this time, Ned was sure that the fellow was up to not much good and probably wanted his beloved tug to transport contraband or for some other shady dealings. But business was business, and what the guy did with the boat after he bought it was his concern. Thus began Ned's version of negotiations. The Jerseyite came up to Essex, and Ned told him his asking price, and also told him it was not negotiable. Immediately, the other fellow made a counteroffer. Ned explained to him what nonnegotiable meant and informed him that every time a counteroffer was proffered the price of the boat would rise by $1,000. This brought the haggling to a quick conclusion, and the new owner of *Chaco* wrote a check and took his boat down the river for the last time and away from Connecticut forever.

But that was not the end of the relationship between Ned and his beloved tugboat. In July 1986, Ned was making his way down Long Island Sound on his way to see the rededication of the Statue of Liberty. To the seasoned tugboatman's amazement, he spied, sitting up on land on City Island, New York, none other than *Chaco*. It was settled forlornly into what appeared to be its final resting place. The astonished veteran skipper immediately pulled over and docked. He made his way up to the tug and clambered aboard. Sure enough, it was his old companion and dearly loved friend. In a moment that must have been full of emotion, he took his rigging knife out of his pocket and pried the brass nameplate from it place in the pilothouse. He now proudly displays it in his Deep River home. It reads: "Essex Boat Works… Essex, Conn…Builders." That day, Ned also discovered the wooden box that once contained the bottle of champagne that christened *Chaco*.

Ned is retired now, but his is a life dedicated to the Connecticut River. In 1958, when he was sixteen, his grandfather loaned him $436 with which to purchase a sixteen-foot Flat Iron Skiff from boat builder Art Finkeldey, who constructed boats on River Road in Deep River. Ned recalls Art's yard as "always a good place to go. Always interesting. Always smelled good from all that wood." Finkeldey christened his creation the *Black Duck*. Ned then bought a ten-horsepower Johnson outboard motor from Essex Paint and Marine on a monthly installment plan, and off he zoomed on a riverman's life. Or, as he and his buddy Strick Hyde like to refer to themselves, a River Rat's life. The boys worked together on *Black Duck* in the spring netting shad. They spent hours on the river just fooling around and having a good time. Strick and Ned continued their river careers together summers as launch boys for the Essex Yacht Club. They cleaned the toilets, waxed the floors and shuttled the "swells" back and forth from their yachts to the shore.

But it was not all work and no play for the high-spirited lads. They enjoyed more than their share of teenaged hijinks on the river. For example, in a burst of enthusiasm, Strick was powering his skiff up the river at a speed much greater than the law allowed, with Ned as a passenger. He was subsequently pulled over by the very first Essex Police Boat. He received a stern lecture that probably went in one ear and out the other. The grumpy officer of the law then proceeded to write Strick a ticket to reinforce his message. While the cop was busy with his pad, Ned took the opportunity to rearrange the stick-on letters on the side of the constable's boat. As a result, the proud upholder of the law took off down the river with the side of his vessel advertising that he was the SEX POLICE!

Ned had his own run ins with the waterborne minions of the law. At one point, he put a twenty-eight-horsepower engine on the *Black Duck*, which gave it the wherewithal to get up and go. He did just that when he gunned it off Essex Island and caught the attention of the now properly signed Essex Police Boat. Ned's rig was much faster than the cops, so he led them on a merry chase around and around a dredge that was anchored in the river. This lasted two hours or so until Ned got bored and snuck away to hide his boat on the riverbank. He leisurely walked back to the Shad Fishing Pier, only to find his pickup truck surrounded by a squad of Essex's finest. Hiding in a spot where he could see them, but they couldn't see him, Ned made himself as comfortable as possible on a hard stone bench and slept off the several beers that he had consumed. The cops eventually got tired of hanging around and left. Ned drove home and woke up with a bit of a hangover and a good story to tell his buddies.

Teenaged Ned wasn't overly fond of formal education, so with his mother's encouragement, he got a job with Holbrook's Connecticut Dock and Dredge Company as a fireman tending to the steam boiler on the *Chaco*'s barge. This proved to be the pathway that led to his life's work. For decades in the twentieth century, Ned could be found up and down the river. He built docks, drove pilings, raised boats and generally did whatever was required of him in order to make a living on the water. While working at the Essex Island Marina, Ned renewed his relationship with Art Finkeldey. Art built the *HoHo*, a mooring pulling barge for his work at Essex Island Marina. Ned and *Hoho* worked at Essex Island building docks, pulling moorings, hauling boats and generally making themselves useful.

At one point, Ned and Strick decided they would forsake the river and start a land-based carpentry business. But, as Ned put it, "we definitely didn't set the world on fire." They quickly returned to the river to continue

Ned Libby (*left*) and Strick Hyde spent decades on the Lower Connecticut River, building docks, driving tugboats. *Courtesy of W. Griswold.*

to seek out their fortunes. For the last nineteen years of his working life, Ned was in business for himself, doing what he loved on the river he loved. He sold this dock-building business in 2005 and now enjoys a well-deserved retirement with his spirited wife, Carol, and their corgi, Cash. He is based in Deep River, but different times of the year can find him at various points between Maine and Florida.

Chaco wasn't the only tugboat built at the yard. In May 1966, according to the *New Era* newspaper,

> *Well over a hundred friends and well wishers were present at the launching of Susan Constant, Nick and Jane Neidlinger's cruising tugboat at the Essex Boat Works. Right on the dot of two o'clock Marjorie Luke, sponsor of the new boat, bashed the traditional bottle of champagne against the cutwater of the boat, and the boat started its slide into the river.*

It floated nicely to its lines, just as it was designed. The tug, built for pleasure, not work, had "many features of unusual comfort." These included "hot and cold running water, a Franklin stove, electric refrigeration, and a gas stove for cooking." Although the *Susan Constant* looked like a no-nonsense workaday boat, it was a yacht, in the true sense of the word, a boat designed for pleasure.

Nick Neidlinger was to play an important role in ameliorating the effects of a tragedy that effectively brought wooden boat building at Essex Boat Works to its sad end. On the night of January 20, 1968, a fire raged through the sheds and grounds of the venerable yacht facility. The cause of the blaze was believed to be faulty wiring in a heater on one of the boats. Three of the yard's sheds and ninety-six "large yachts" were destroyed in the conflagration. The extent of the damage was maximized because most of the boats were stored with full fuel tanks to prevent damage caused by moisture condensation over the winter.

Neidlinger, who had a special relationship with the Essex Boat Works since his recreational tugboat was built there, decided that he would do whatever he could to help the yard, and especially the guys who made their living there. Many of the employees lost the tools of their trade and equipment, including clothing, in the fire. None of it was insured. Nick knew these men well, and in response to their loss, he created a fund to help. He initially sweetened the pot, and community members and friends of the boatyard chipped in to help. They were able to generate enough money to give the workers interest-free loans to replace the implements destroyed in the flames. As the Essex Boat Works was rebuilt, its crew was well equipped and ready to go to reestablish the yard in a bigger and better iteration. As the men were able to repay the loans, the money was donated to the Essex Fire Department in recognition of the invaluable service it provides to the community.

Ned recalled the seventy-two-foot steel lobster boat built by the Essex Boat Works, the *Mark Derren*, which was launched in the spring of 1977. As of this writing, it is still plying the crustacean trapping trade in the waters of Rhode Island Sound. At the time *Mark Derren* was constructed, Essex Boat Works was owned by a popular Essex denizen named Stu Ingersoll. Stu was a firm believer that boats, be they for work or pleasure, should be beautifully turned out and appointed. He was a true perfectionist. As a result, any profit that could have resulted from building the *Mark Derren* were spent, in Ned's words, "makin' her too fancy. She shone like a mirror, and her cabin looked like a damn yacht." It was the first large boat built at the yard since the disastrous fire. A *Hartford Courant* article at the time called it "one of the largest boats built on the Connecticut River in the past 100 years."

The major difference between the *Mark Derren* and all the Essex ships and boats that preceded it was that this new lobster catcher was made of steel. The article says that Essex Boat Works has a "substantial reputation for its craftsmanship involving wooden boats." That craftsmanship served the yard well in its new creation—even though the hull was steel. Below decks, it had "oiled teak panelling over heavy insulation batting in the fo'c's'le and galley area." The reporter breathlessly announced that the boat would feature an "electric microwave oven," an exotic innovation at that time.

But the high-quality appointments were evident above decks as well. "Touches of luxury extend also to the bridge and control station with Burma teak over insulation to both deaden the sound of the 350 horse power diesel engine and absorb sweating from the quarter-inch steel hull." Much to the amazement of the reporter, "FM stereophonic music...piped onto the working platform, will be an added touch." All those touches combined with Stu Ingersoll's insistence on a yacht-like product gave the captain and crew of the *Mark Derren* a comfortable workspace for their five-day offshore trips eleven months a year. It could cruise at eleven knots, speed being important to avoid bad weather and get the catch to market as quickly as possible to ensure the best price for the lobsters.

The *Mark Derren* was the last boat of size that was built top to bottom at Essex Boat Works. The yard continued its tradition of fine craftsmanship finishing and fitting out boats whose hulls were put up at other yards. An example of this is the Essex 30 commissioned by Ted Lahey, the owner of the Essex Boat Works in the late 1990s. Her "classic Downeast hull" was built in Maine by the Young brothers. The deck and console, however, were constructed in Essex. The center console was raised to accommodate the engine and improve the sight lines of the driver. The array of electronics is flush-mounted into the console. The raised forward deck makes an excellent casting platform for fishing and provides a little headroom in the cuddy cabin below. The cabin features a V-berth, a head and a hanging locker. *Yachting Magazine* gushed that the versatile design "is just the ticket for the yachtsman who enjoys a variety of waterborne activities."

Today, the Essex Boat Works "remains dedicated to the on-going service and storage of the region's finest boats and yachts....An experienced crew is on staff to handle all boat constructions, including, wood, fiberglass, composite, aluminum and steel....Knowledgeable yachtsmen choose Essex Boat Works to maintain and store their boats."

Seth Perrson

If a ninth-century male Viking found himself teleported to the banks of the Connecticut River on the 1950s, he might well have looked just like Seth Perrson. This mythical time traveler would have shared a love of boats and the sea with Seth as well. A second-generation Swedish son of a boat-building father, Perrson became disenchanted with formal education in the eighth grade and decided that he would dedicate his intellect and energy to the building of boats. It was a wise decision. He built his first, a small sailboat, at the age of fourteen in the basement of his family's row house in Brooklyn, New York. His initial build was a twelve-foot sloop of his own design. He was encouraged by his father, who also built small boats in that basement. His dad was a true blue-water sailorman who made several passages around Cape Horn. He built several sail and powerboats, including one that required him to dismantle part of his house to get it out of the basement. At one point, he designed a boat specifically for laying telephone cable. The apple didn't fall far from the tree in the Perrson family. Seth followed his father, Frans, as a builder of wooden boats.

At seventeen, Seth went to work for Consolidated Shipbuilding Company. He rapidly rose to a position of importance in the operation, leapfrogging over men who were considered to be master craftsmen. His innate talents, coupled with ambition and energy, singled him out as a candidate for advancement in the industry. While at Consolidated, he built his second boat. It was based off a do-it-yourself article in a magazine and turned out to be a handsome seventeen-foot sloop. Before his twentieth birthday, Seth left that company and struck out on his own. He opened a boat-building shop in his hometown of Brooklyn. Work quickly found him via word of mouth. His first customer happened to be a distant relative of George Washington who commissioned Seth to put together a twenty-eight-foot powerboat, and Perrson Boats was up and running.

The young maritime craftsman then began to exclusively build boats from his own designs. He made eighteen- and nineteen-foot sloops with a sail design that allowed them to race at area yacht clubs. His boats proved to be fast and reliable. In several competitions, he gave boats that were designed by L. Francis Herreshoff (the generally acknowledged grand master of the yacht design world) a close run for their money. This enhanced his growing reputation as someone to be reckoned with on the boat-building horizon. As he developed confidence in his own abilities and optimism about the third decade of the twentieth century, the youthful builder began to generate some interest in East Coast boating circles.

But, as it did for so many, the Great Depression did its best to derail Seth's projected path to the future. Suddenly, clients no longer had the money to pay for boats that were ordered and then built. Inventory began to encroach on workspace. New orders slowed to a trickle and then basically dried up. Frans, the family patriarch of Cape Horn fame, found himself compelled to retire from his job with the telephone company for which he had designed the innovative, cable-laying boat. It was a difficult time for the family, as it was for so many people throughout the world whose lives had been crushed by the economy's crash.

The Perrsons decided to respond to changing economic landscape and migrate to a new location. They didn't go too far, only one hundred or so miles east up the coast to Old Saybrook, Connecticut. This move positioned them ideally right between Boston and New York. It gave them access to these very large markets and placed them under the Baldwin Bridge on the banks of the Connecticut River. The location afforded easy access to the vibrant estuary's boating communities, as well as those up, down and across Long Island Sound. In 1931, Seth and Frans began to build their shop and yard on a two-and-a-half-acre piece of land. Father and son were familiar with the area because they frequently visited members of their extended family who lived along the historic river.

While Frans did much of the actual construction work on the boat yard buildings, Seth picked up jobs at the Jakobson Shipyard back in New York. He was the loftsman, responsible for turning the marine architect's design into an actual boat. He was a key player in the building of the last wooden tugboat and the first steel tug that the yard built. At only twenty-three, Seth Perrson carried a tremendous amount of responsibility. Fortunately, he rose to the challenges presented to him. The three-dimensional pieces that resulted from his drawings fit so precisely that a grizzled old German welder who built boats for the German navy in World War I acknowledged that "even the U-boats never had fits so tight." Jakobson placed a lot of confidence in the young man, and the company was rewarded with some excellent pieces of work. But Seth was anxious to strike out on his own, and as soon as Frans had the shop ready to go, he migrated permanently up to Old Saybrook to begin building his own boats.

Seth and Frans's first paying job was to build the spars to refit onto an 1857 sailing vessel, a cutter named *Viola* (also the name of an Ernie Way–designed rumrunner). Frans was particularly adept at this chore; his years working for the phone company gave him plenty of experience splicing and braiding wire. After *Viola* was remasted, Seth put together five one-design racing day sailers that were named the Old Lyme class. These early boat-building forays gave the shy lad pause to think. Would he prefer to be a designer? A builder? Both? He pondered his future and came to the ironclad decision that he would be a builder. And not just a builder, but one of the best builders in the business.

Once that vector of his career path was settled, Seth cultivated an excellent working relationship with marine architect Winthrop "Wink" Warner. Warner is a legend along the Connecticut River and beyond. He designed everything from dinghies to draggers. His oeuvre included motor-sailers and motorboats (and even the Chester-Hadlyme Ferry), but his specialty was cruising sailboats. Seth built the Warner 20 cruising sloops *Sea Horse* and *Diablo*. They were followed by the thirty-plus-foot sloops *Snapper Blue* and *Yankee Girl*. Seth was inspired to go a bit artistic by the *Sea Horse*, and he cast a set of bronze seahorses to decorate the sloop's bow. Handmade ornamental touches and custom fittings became signature additions to his work and made his creations all the more special.

Seth and Frans added a crane and a marine railway to their yard. While it was their prime objective to build boats, they realized that in order to make the operation economically feasible, they needed to maximize potential revenue streams. They added hauling, storage and repair to the services offered. And the Perrson Boatyard became a fixture on the lower Connecticut River for much of the middle of the twentieth century. The gray barn by the river was known to be the birthplace of some beautiful boats.

Seth Perrson's sailboats were lauded for their craftsmanship. His boats won many important races and could also cruise comfortably. *Courtesy of Caryn B. Davis.*

With the dark clouds of World War II on the horizon, Seth knew he would be drafted into military service. In order to keep the yard financially afloat during his absence, he took a job at an area company that built glider airplanes for the U.S. Army Air Force. The talented fabricator was quickly promoted to shop foreman. But his inevitable draft notice soon sent him to India as an airplane mechanic. Ironically, it was the noncombatant Frans who would be killed while the war raged across the globe. The Perrson family patriarch suffered a fatal injury when he fell from a boat that was up on a cradle.

Seth returned from the global conflict resolved to continue the family legacy of boat building. He began to turn out a long string of excellent boats. His high-quality creations took many honors on the racing circuit and gave their owners decades of pleasure and adventure. Among them was a Fenwick Williams–designed eighteen-foot catboat named *Tabby*. After building a thirty-eight-foot powerboat of his own design and a forty-two-foot schooner designed by Murray Peterson, Seth was tasked with building a thirty-seven-foot Herreshoff design sloop. It was named *Rogue* and wound up on the cover of two of the most popular sailing magazines at the time. *Rogue*'s craftsmanship was deemed to be of the very highest standard, and it still races competitively well into the twenty-first century.

Perhaps Seth Perrson's most famous (and most beautiful) creation was *Finisterre.* It was built for Carelton Mitchell, one of the foremost ocean racers of his period. Mitchell was drawn to the Perrson yard by its strong and growing reputation for craftsmanship, attention to detail and quality of construction. The boat of Mitchell's dreams was almost impossible to imagine. He wanted a comfortable cruising boat that could gunkhole its way into shallow water, *and* he wanted it to be a world-class racing boat. Olin Stephen, the genius selected to design this seeming contradiction in terms, stayed awake nights, trying to figure out how to competently fulfill his commission. He finally allowed himself to get a good night's sleep, when he realized that Seth Perrson had the imagination and skill to build a boat that wouldn't fall apart in a hurricane, could be easily sailed single-handed, win races and sleep seven comfortably.

Finisterre was a racing phenom from the get-go. Despite such luxurious appointments as a fireplace, a stereo, a refrigerator, scads of electronics and two generators, it was wicked fast and seaworthy. *Finisterre* asserted its racing dominance by winning three Newport to Bermuda races in the period from 1956 to 1960. It won the Miami–Nassau race twice during that time and took the gun (and the fun) in many other important contests of the period. Well into its seventh decade today, *Finisterre* is still winning races and exploring the harbors of Europe and the Adriatic. It remains a testament to Olin Stephen's faith in Seth Perrson.

Finisterre cemented Seth's reputation as a first-class boat builder. Requests for his creations piled up, and he proceeded to produce some of the finest yachts of the period. These included the thirty-eight-foot John Alden ketch *Abigail*; *Fox Fire*, a Dunham and Timken keel/centerboard forty-footer; and *HiQII*, which at forty-five long was one of the largest sailing boats built at the Perrson yard. It was followed by *Easterly*, a Sparkman and Stephens design, built for Richard Cooper of New Britain, Connecticut. Cooper's charge to Perrson was for a boat "with no straight lines." It is believed to be Seth's personal favorite among all the boats he built. Everyone who has sailed or worked on it agree that *Easterly* is truly magical in its form and function.

In June 1964, a truly terrible tragedy was visited upon the Perrson Boatyard. Even more devastating, this totally unnecessary tragedy was caused by humans. For reasons that remain obscure, arsonists torched a boat that was on the ways ready to be launched. The resulting conflagration destroyed the entire Perrson Boatyard complex. Ten other boats, all Seth's tools and papers, his designs, images—everything went up in an evil conflagration. Among the most disheartening parts of this sad story was the loss of all his father's memorabilia from his seafaring days around Cape Horn. His well-worn sea chest, laden with the dreams and adventures of a lifetime—up in smoke. Seth was brokenhearted, but not broken.

Immediately after the fire, Dauntless Club member Bill Slaymaker, the customer for whom the about-to-be-launched burned boat was being built, ordered another one. Community members, friends and supporters rallied around Seth in his time of need. His network saw to it that he had the financial and logistical support necessary to rise from the ashes and continue to do his wonderful work. It took a few months, but the yard became functional once again. Appropriately, Slaymaker's new boat was named *Phoenix*.

More splendid boats were to follow. *Ann Toy*, a thirty-seven yawl; a party fishing boat; a Connecticut River shad boat; and several sloops were launched from the yard. Over some time, Seth built a Herreshoff 25 as his personal watercraft. It took two and a half years to build because he was always busy with someone else's boat. Once completed, it allowed him to experience some of the joy that his creations brought to his customers throughout the years. As Seth began to wind down a bit, his sons Rick and Jon took a more active role in the yard. They started to build boats on their own to carry on the family tradition in the wake of Frans.

Seth died in June 1980. His passing marked yet another significant moment in the history of wooden boat building on the Connecticut River. Some of his lovely creations are still sailing today, and may they do so for decades to come. His sons continued the business but eventually scaled back to make rowboats, bass boats, kayaks and dinghies. The business eventually moved from Old Saybrook to Centerbrook. It is still going strong as of this writing. Would-be boatwrights can go there to learn how to build small boats from scratch. The shop does repair work and also builds boats from kits. It is meet and right that the legacy of Frans and Seth Perrson still resonates from Cape Horn to the Connecticut River.

Earle Brockway

The Brockway family has lived on the lower Connecticut River for over three hundred years. Brockway Island sits off Great Meadow Road in Essex. Both Lyme and Deep River have Brockway Ferry Roads that travelers used when Brockways poled, rowed or sailed them across the river on flat-bottomed scows. A section of the river is known as Brockway's Reach. Generations of family members built sloops, schooners and brigs during the eighteenth and nineteenth centuries. The Brockways were known for the fast privateers they launched to prey on British ships. The family legend was carried on by a singular scion of that shipbuilding family, and his name is reverently associated with thousands of unique vessels. His designs are still being built from Maine down to the Chesapeake and beyond. In fact, they are still in production all over the world. This remarkable individual created no-frills, totally utilitarian, workboat types that are ideal for shadding, clamming, oystering, inshore fishing or just messing around on the water. His boats, and those his work inspired, are simply called Brockways. They are prized among folks who love to work and play in creeks, rivers, bays, sounds and oceans.

This unique craftsman was born in Moodus, Connecticut, in 1918 or so. The exact dates of his birth and demise are flexible. His given name was Richard, the same as his father's. But he quickly became known to one and all by the noble sobriquet Earle. Earle lived up to the royal roots of his name; he became a high-ranking member of the boat-building nobility that evolved along the lower Connecticut in the twentieth century. Many consider him to be the most prolific boat builder in American history. Like a proper nobleman, he combined eccentricity with purpose and proudly represented his lineage of shipbuilding progenitors. His story is rife with anecdotes of those who knew him. All the stories told about him attest to his extreme individualism, his work ethic and his humble, yet splendid, creations. His boat-building story began at the onset of the Great Depression, and economics played a key role in shaping his destiny.

The crashed and tattered economy of the early 1930s forced Earle's father to close his small machine shop in the Moodus section of Connecticut, an area famed for its mysterious noises and demonic legends. Richard Sr. borrowed some money, packed up his growing family and moved them to Old Saybrook, on the western shore of the mouth of the Connecticut River. He then proceeded to follow his family heritage and started up a boat-building business. Earle's dad decided to continue the Brockway tradition of building sailing vessels. He fashioned elegant thirty-foot sailboats out of traditional woods like pine, cedar and oak. Young Earle soon became entranced with any and everything that had to do with boats. His father cobbled an eight-foot rowboat together, and the eager boy would jury rig a feedbag to an oar in an attempt to sail it. Dad then cut and shaped a cedar pole into a mast, his mother stitched him up a sail and off the lad went to explore the creeks and coves of the river and sound. Thus began a lifelong obsession with small craft.

When he wasn't out exploring and adventuring, Earle could be found at his father's side, absorbing knowledge about the ins and outs of putting a boat together from scratch. His father was a jack-of-all trades. He was a machinist, an engineer, a carpenter, a mechanic and, later on, a town selectman. But his passion, which he passed on to his son, was building boats. Dad learned many of his skills from Joe Lord, a scion of another old Connecticut River family who lived in East Haddam. After creating a few sailboats, which were difficult to sell in a depressed economy, he became primarily devoted to building powerboats, rather than sailing vessels. His boats were often thirty feet long or larger and featured thick oak keels, strong planks and sturdy, heavy ribs. He was able to school his son in the essential basics of boat building and imparted an important sense of the quality that goes into making a good boat something special. Young Earle absorbed this wisdom. He was grateful for the lessons and carried them with him throughout his life.

After his graduation from Old Saybrook High School, he worked at a couple of local boatyards, honing his skills on small sailboats. He was of draft age during World War II and was conscripted into the army as it cycled into postwar peacetime. Earle returned to Old Saybrook for good in 1947 and moved into the Floral Park section of town at the end of Fourth Street. His goal was to build high-quality sailboats, like his father. Both men believed that a well-built boat should last for decades and took great pains to use the finest materials and careful craftsmanship to ensure that their products would be around for a good, long time. He had gained some early experience in local boatyards, building Star, Lightning and Comet class sailboats. In 1948, he struck out on his own. He put up a little sign that read Brockway Boatworks on his Old Saybrook property, and off on his career path he went. (A few years later, the sign was knocked down by a truck. He never bothered to replace it.) For openers, he built a splendid thirty-foot sloop that was to become a critical lesson in supply and demand. What he learned from that experience motivated him to think that selling a lot of inexpensive boats might be more lucrative than selling (or not selling) more costly units. That decision shaped his future and created an important niche in Connecticut River boat-building history.

Even as the postwar economy began to percolate, building well-turned-out sailboats with first-rate materials proved not to be a cost-effective business model. It was labor intensive work, and the finest wood was expensive. So were sails. Earle purchased Egyptian cotton and sewed his own. Several boat builders in the area were squeezed by these factors and reluctantly shuttered their sheds. Earl didn't have much use for a shed though, even when he had one. He preferred to work outdoors in all weather, usually lightly dressed. He allowed that his work kept him warm and that being indoors made him a bit claustrophobic. His plein air preference cut down on overhead a bit, but he still found it difficult, if not impossible, to turn a profit with his high-quality sailboats. When his one shed burned down in a disastrous fire, he never bothered to rebuild it. His beautiful sailing boats sometimes sat in his yard for quite a while before a discerning buyer would see their value and purchase one.

Faced with the grim possibility of going out of business, Earle decided to get innovative and try to adapt to a changing market and explore new technologies and techniques. The escalating cost of traditional oak and cedar planking and the time-consuming method of bending it around a prepared frame needed to be looked at from different angles to see if alternatives were possible. Earle set his mind to creatively solve the conundrum confronting him. Given the variables he had to deal with, the young entrepreneur in an old business came up with a brilliant solution—plywood! While, perhaps, not as aesthetically pleasing as cedar, pine and oak, plywood was, above all else, inexpensive. Earl quickly turned his intuition into a series of steps that allowed him to ultimately build thousands of boats, some of which still ply America's waters today, though their numbers are diminishing. His boats and designs still inspire the adulation of legions of devotees and admirers.

The insight that set him on this path was the realization that with plywood, he could shape the boat and then add the internal braces and supports. As he perfected this construction method, his boats became increasingly strong, seaworthy and, importantly, economical. He began to sell his utilitarian craft to local fishermen and clammers. There was also a growing market for small rowboats that vacationers could enjoy at their summer cottages. The development of small fiberglass boats cut into the tourist trade eventually, but his workboats became the standard for those in search of shad, lobsters, oysters and other delectable creatures from the river and Sound. While he may have longed to construct exquisite yachts, the meat and potatoes reality of supply and demand dictated that he create quotidian craft that made up for their lack of traditional loveliness with an all-around function of usefulness.

Earl settled on two basic designs: sharp-bowed skiffs and blunt-bowed scows. He also made rowboats in the eight-to-twelve-foot range and the occasional sailboat, but his bread and butter were the skiff and the scow. When his brother and father worked with him at the yard, he was able to turn out upward of three hundred boats a year. When he labored solo, his output was in the range of fifty boats a year. Still, that is one a week. He was able to keep up that output because he was truly indefatigable. He could be found in his yard 365 days a year often for ten or twelve hours a day. His products worked the waters up and down the Eastern Seaboard. Cape Cod was home to hundreds of Brockways. In Rhode Island, they were used for bullraking clams from the mucky bottoms. In Connecticut, they were used for gillnetting shad on their annual runs up the river in April and May. They were used for lobstering, oystering and scalloping, too. They became staple of marina rent-a-boat fleets, due to their modest cost and the fact that they were virtually indestructible. Their flat-bottomed design make them tremendously stable work platforms from which to haul edible animals from the depths. Clammers could easily haul a four-hundred-pound dredge over the gunwale of one of his scows.

The other feature that added to the appeal of Earle's boats was their modest price. As late as the middle 1980s a sixteen-foot skiff would set you back only $700 or so. A twenty-foot scow was a pocketbook-friendly $1,200, and for the more economy minded, you could pick up a nine-foot skiff for only $300. No wonder hundreds of customers found their way to Earle's yard to pick up a boat that would serve them in good stead for years. Finding Earle wasn't that easy. He never advertised. He didn't need to. Word of mouth sufficed to keep a steady flow of eager buyers streaming into his Floral Park fiefdom. The main reasons his boats were so inexpensive was that he was the primary source of labor and the materials he used were available from local lumberyards and hardware stores. In the early days, as his boats became more popular, he began to offer a delivery service and could be seen towing a string of skiffs and scows up and down the sound and over to Long Island. As demand grew, however, his work schedule dictated that would-be buyers come to him.

His boats were crafted from one-half-to-three-quarter-inch exterior plywood and fir or pine stock for chines, stems, gunwales and keels. Roofing and finishing nails, bolts, glue and tar hold the boats together. Earle was fond of liberally laying on tar to seal his panels. Since most of his customers preferred to paint their own boats, the standard joke was to plan two days to remove the excess tar and two days to paint. Earle would provide a painted product if the customer requested it. He did most of his work with everyday hand tools. He felt that they gave him more control over his finished product and were a heck of lot a quieter than grinding motors and power saws. He would pound pegs into the ground and use them to shape wood into the outlines of his boats. As time went on, though he relied on jigsaws and skill saws a bit more.

Those customers who made their way to his boatyard found themselves in a somewhat alternate reality, where boats were hauled across the ground by lines hooked on the door handles of beat-up luxury automobiles. Oldtimers still refer to his "Packard Period" or his "Cadillac Period" when distinguishing which boats came from what era. Today, he might have been referred to as a hoarder, but to Earle, everything had a potential use in the future, so nothing should ever be thrown away. The result was a jumble of junked cars, car parts, broken tools and just lots and lots of stuff. Earle's frugality led him to say that the only thing left over after he completed building a boat was sawdust. He used rocks and old car parts to weigh down pieces of wood to get rid of a warp or bend to his will. There was always a piece of wire or metal that could come in handy at just the right moment. His living quarters have been described as a "hovel." To say he lived simply would be an overstatement.

For example, consider the folklore that surrounds his Cadillac. Even someone as minimalistic as Earle would occasionally have to drive into town for some basic victuals and supplies. His method of doing so became legend in the annals of Old Saybrook. His chariot was an ancient Cadillac sedan. It was totally rusted out to the point where it lacked basic body parts. It was missing most of its glass. It had no brakes. It was without a gas tank! Earle rigged up a six-gallon plastic marine gas tank on the roof and used gravity to supply his clunker with fuel. One day, Earle was forced to stop on a hill in traffic, which caused his fuel supply to dry up. Stuck in park with no brakes, he was in a bit of a pickle. The local cops, who had tolerated his antics for years, finally said enough. He had no license, no registration, no plates, no insurance and no mufflers. His sweet Caddy was towed back to his boatyard, where it lived out its days moving boats around.

Pesky motor vehicle regulations were not the only legal strictures that were to become bothers to Earle. As more and more people began to own boats and spend time on the water, the U.S. Coast Guard began to be concerned with things like boating safety. Officials started to demand that he register as a boat builder and carve hull numbers into his prized creations. Then came the edict that he add upright and level flotation to his boats. Buyers thought this was absurd and would rip out the stuff just as soon as they bought their boast. Then came bureaucrats snooping about, nattering on about old Earle registering as a business and (gasp!) paying taxes. They proceeded to do all the things robotic state suits can do to make life hell for a small business owner. The world in which Earle could truly be his own man fading into cyberspace. He adapted as best he could, but he didn't like it very much.

But in terms of personal style and his work environment, Earle remained a one-of-a-kind crusty old Yankee. He was the epitome of self-sufficiency. He worked all day outdoors in any weather. His bald head was hatless, and his skinny frame was clad only in a thin jacket, even when the thermometer hovered near negative numbers. Personal hygiene was never his strong point. His self-reliance was so ingrained that when his boatyard was being engulfed by a raging fire, he chose to drive a mile to the local Howard Johnson's to call the fire department rather than bother his neighbors in the middle of the night. Of course, he didn't have a phone of his own. The fire, though quite devastating, proved to be only a minor setback. The simplicity of Earle's operation allowed him to get back to business within a short period of time. His sturdy boats continued to be put together in Floral Park.

The uniqueness of his operation and Earle's one-of-a-kind personality often drew the curious, the interested and potential customers to just hang

Classic Earle Brockway skiff painted like the Cornfield Point Lightship. *Photo by Liz Brunell.*

out at his boatyard to be amazed and soak in Earle's pithy wisdom and aura. One such fellow was a young man named Greg Myerson from down the sound a bit. He spent three seasons trapping muskrats in order to come up with the $700 needed to purchase his Brockway. He spent a few days with Earle, watching him create the boat that would take its place in angling history. Greg was an avid fisherman. On October 19, 2011, he was trolling along in his Brockway off the coast of Branford, Connecticut. Something big smashed into his lure. When the battle was over, Greg became the holder of the new world record for striped bass with an 81.88-pound trophy to proclaim his preeminence. He credits his Brockway for not only providing a fishing platform but also bringing him luck.

In a true testament to the validity of Earle Brockway's life work, it is most fitting that his designs have reached far beyond the Connecticut River and the East Coast of the United States. Timothy Visel, a key founder of the Sound School in New Haven and a longtime New England waterman, purchased his first Brockway in partnership with his brother in 1972. He fell in love with the design: "The boat was tough, could take a lot of punishment from the gear and was a good sea boat. Most importantly it was reliable." In response to an international request for a small, simply constructed, strong plywood workboat, as part of the United Nations' response to floods in Asia, he submitted the Brockway design. The response was such that the work of

Earle Brockway design would soon become available worldwide. They have been spotted as far away as Fiji and are common throughout Asia. Earle's spirit continues to put practical boats on the waters of the world.

Typically, Earle didn't ask anything in return for his handiwork. When offered compensation, he responded, "Commercial fishermen need a help once in awhile. I just hope they will come and visit someday." And that is exactly what happened. A few years before Earle died, a visitor from Haiti came to see him with photographs that showed several Brockways gathered on a Haitian beach. He had come to thank the man who had done so much to help his village. It was a touching meeting between members of the Brotherhood of the Sea. As years have passed, the occasional builder claims to make Brockways. This raises some hackles in the boatbuilding community. The sentiment is that anyone can build a Brockway design, but only Earle could truly build a Brockway.

Earle Brockway worked at the craft he loved up to the very end of his life. A day or two before his death at age seventy-six (or seventy-eight depending on the source), a neighbor spied him moving a twenty-six-foot boat down to the river. The process involved putting rollers under the bottom and hauling it along with his Cadillac. The boat made it to the water. Earle soon made it to the Great Beyond. Local lore has it the country's most prolific boatbuilder died doing what he so singularly loved—setting up a skiff, with a hammer in his hand. Shortly before the end, Earle expressed the wish on more than one occasion that "when all is said and done, and my last boat has been built, I'd like to be launched in a Brockway, like a proud Viking, and put to rest at sea."

Art Finkeldey

If you were standing at the Roger Tory Peterson Boat Launch on the Back River in Old Lyme on a summer afternoon, there would be a bustle of boaters putting their canoes, kayaks, fishing boats and day sailers. After they jostle for space, become waterborne, get paddles in the water, motors started and sails up, most of them head downstream. Just below the launch, swinging at a mooring, is a beautiful eighteen-and-a-half-foot sloop named *Saloola*. It is a Charles Mower–designed Petrel class. *Saloola* was built a few miles up the Connecticut River in Deep River by yet another legendary Nutmeg State boatwright. His name was Art Finkeldey. Some of his wooden boats are still afloat many decades after he put them together in the 1950s and '60s.

His workplace was located on River Road, between Deep River and Essex. The neighborhood today is dominated by the waterfront property of the very well-to-do, but Art's unassuming nineteenth-century house still sits by the river. The old-timers who were fortunate enough to step foot in his workspace and soak in his wisdom usually begin their reminiscence by recalling how good Art's boat shop smelled. They recall the mingled aromas of oakum, pitch, cedar, oak, pipe tobacco, red lead paint, wood smoke from a pot-bellied stove and coffee. Art was a genial man; he enjoyed a good conversation but remained focused on the task at hand with singular purpose and devotion. He would always be steaming frames or planing a piece of wood to perfection. His steadfastness resulted in many splendid boats.

That smell is the first thing Ned Libby recalled about his time spent with Art and his family. Art built Ned's first boat. Ned remembers the christening ceremony for the *Black Duck*, which was arranged and presided over by Art's daughter Ann. She smashed the requisite bottle of champagne on the freshly painted bow and the sixteen-foot Flat Iron skiff slid down the rails of Art's makeshift railway and came to rest exactly on its lines in the Connecticut River. This ceremony reinforced in the teenager's mind the importance of ritual in maritime matters. His work as a launch boy at the Essex Yacht Club had a ritual component too, as he was responsible for raising and lowering the flag for "colors" on a daily basis.

Black Duck wasn't the only vessel that Art Finkeldey built for Ned Libby. A few years down the river, he put together the *HoHo*, a small barge. Ned used the *HoHo* for almost twenty years. He worked it to build docks, repair docks, move docks and drive pilings. He also set and pulled moorings, moved equipment and performed dock maintenance. *HoHo* provided Ned the opportunity to continue to make his livelihood on the river he loves. It was also one of the last barges built on the Connecticut, which had a long history of barge-building and usage.

Art also built Blue Jay–class day sailers. He put together a special one for his daughter Ann. These thirteen-and-a-half-foot little sloops were a mainstay of yacht club racing fleets for decades. Ann made quite a name for herself at the Pettipaug Yacht Club, just down the river from her dad's shop. She consistently vanquished her frustrated foes in fiercely contested and observed races. In the small sailboat department, Art gained a bit of renown when marine architect William Atkin named one of his designs after him. Plan for the Finkel-Dink are still available today. It is a nine-foot bluff-bowed pram with a spritsail.

Art's marine railway allowed him to haul fifteen or twenty boats for winter storage, and he did repair work. But putting boats together from scratch was his true passion. His sailboat output included an impressive H-28 ketch. Designed by Herreshoff, it was a full-keeled cruising boat capable of passage-making in style and comfort. The aforementioned Mower-designed *Saloola* was another signature sailboat from his yard. It was actually a replacement boat. Shortly after the purchasers took delivery of *Saloola*'s predecessor, *Flash* unfortunately burned. The owners were so impressed with Art's workmanship, however, they immediately placed an order for a new one. They decided the name *Flash* was a bit too combustible and settled on *Saloola*, which is an amalgamation of a gal named Sal and the Swedish pronunciation of Olaf.

Saloola, built in 1962 by Art Finkeldey and lovingly maintained by Frederick Schavoir, in its slip at the Essex Boat Works. *Photo by Liz Brunell.*

Today, Art Finkeldey's legacy is encapsulated in the remaining twenty-four and twenty-six-foot custom bass boats that he built from his own designs. Mystic boatwright Jeff Hall speaks lovingly about them. He should know—he owned one for years. He praises them as "good sea boats and good looking, too." He keeps track of the three that he knows still exist. *Gull*, *Scaup* and *Oyster* can be found in Long Island Sound and the Connecticut River to this day. Art built a twenty-four-footer as his personal boat. It was a combination work and pleasure craft. He used it to go lobstering and to take his family on pleasant excursions up and down the river. Wesleyan University purchased two of Art's twenty-four-foot power boats to use as chase boats for its crew teams upriver in Middletown.

The remaining Finkeldey-built boats are a testament to the skill, time, effort, energy, money and love requisite to keep these wonderful creations afloat so that future generations might enjoy and appreciate them. Old-timers, like Art, used to be fussy about who they built boats for (when economics allowed them to pick and choose). But boats get passed on to family members or sold outside the family. Sometimes, a fine boat will wind up in the hands of someone who may not treat their splendid craft as they should.

An example of this is related by a Southeastern Connecticut boat builder/restorer. The scion of a distilling fortune purchased a Finkeldey twenty-foot launch. He had the builder/restorer put it into tip-top shape. In company with another boat, they took it out on the Mystic River for sea trials. The new owner kept paying rapt attention to his new rpm calibration unit and ignored the warning of the other boat to get back into the channel when, *crunch*. His newly painted hull met some unforgiving gravel off Sixpenny Island. But that was not the worst of it. The new owner had "invented" what he called an "inverter/converter" that was supposed to prevent electrolysis. Unfortunately, it *caused* massive amounts of what it was supposed to prevent. The result was a boat that was rapidly turning into "shredded wheat." The builder/restorer put it back together and could have guaranteed himself work in perpetuity if he left the unit alone. The thought of working for the scion long term was not one to be entertained.

Art Finkeldey was the last prolific commercial builder of wooden boats on the lower Connecticut River. One or two hobbyists may turn out a boat or two every now and then, but the centuries-old tradition of turning trees into boats is, for all intents and purposes, over. Fiberglass and composites now make up most of the boats on the water. They are fabricating aluminum skiffs, based on Earle Brockway's designs, up at the Chester Marina. It is wonderful that a dedicated cadre of owners and craftsmen, with a love of the beauty that is a wooden boat, still put the time, energy and money into these excellent examples of human ingenuity that will keep them on the water.

Selected Sources

Albion, Robert Greenlaw. *Square-riggers on Schedule*. Princeton, NJ: Princeton University Press, 1938.

Beers, J.H. *The History of Middlesex County, 1635–1885*. New York: J.H. Beers & Company, 1884.

Buker, George E. *Blockaders, Refugees and Contrabands: Civil War on Florida's Gulf Coast, 1861–1865*. Tuscaloosa: University of Alabama Press, 1993.

Canney, Donald L. *Lincoln's Navy: The Ships, Men and Organization, 1861–65*. Annapolis, MD: Naval Institute Press, 1998.

Carrick, Robert, and Richard Henderson. *John Alden and His Yacht Designs*. Camden, ME: International Marine Publishing, 1983.

Clark, Arthur H. *The Clipper Ship Era, 1843–1869*. New York: G.P. Putnam & Sons, 1910.

Cutler, Carl C. *Queens of the Western Ocean: The Story of America's Mail and Passenger Sailing Lines*. Annapolis, MD: United States Naval Institute, 1961.

Dana, R.H., Jr. *Two Years Before the Mast*. New York: A.L. Burt, 1940.

Dart, Margaret. *Yankee Traders at Sea and Ashore*. New York: William Frederic Press, 1964.

The Dauntless Club Centennial History 1905–2005. N.p.: privately published, 2006.

Fairburn, William Armstrong. *Merchant Sail*. Vol. 5. Edited by Ethel M. Ritchie. Center Lovell, ME: Fairburn Marine Educational Foundation, 1945–1955.

Goldman, Matthew. *Journal of a Constant Waterman*. Halcottsville, NY: Breakaway Books, 2007.

Grant, Ellsworth S. *"Thar She Goes."* Essex, CT: Greenwich Publishing Group, 2000.

Greenville, Basil. *Evolution of the Wooden Ship*. Caldwell, NJ: Blackburn Press, 1988.

Griswold, Wick. *Griswold Point: History from the Mouth of the Connecticut River*. Charleston, SC: The History Press, 2014.

———. *A History of the Connecticut River*. Charleston, SC: The History Press, 2012.

Howland, Southworth Allen. *Steamboat Disasters and Railroad Accidents in the United States*. Worcester, MA: Dorr, Howland & Company, 1840.

Jefferson, Sam. *Gordon Bennet and the First Yacht Race across the Atlantic.* London: Bloomsbury Publishing, 2016.

Lincoln, Charles H. *Naval Records of the American Revolution, 1775–1788*. Washington, D.C.: Government Printing Office, 1906.

Lloyd, Robin. *Rough Passage to London*. Lanham, MD: Sheridan House, 2013.

Locke, Joseph L., and Ben Wright, eds. *The American Yawp: A Massively Collaborative Open U.S. History Textbook*, vol. 1, *To 1877*. Stanford, CA: Stanford University Press, 2019.

Lubbock, Basil. *The Western Ocean Packets*. Glasgow: James Brown & Son Publishers, 1925.

Malcarne, Donald. *Houses of Essex*. Vol. 1. Ivoryton, CT: Ivoryton Library Association, 2004.

Malcarne, Don, Edith DeForest and Robbi Storms. *Deep River and Ivoryton.* Charleston, SC: Arcadia Publishing, 2002.

Malcarne, Don, and Robbi Storms. *Around Essex: Elephants and River Gods*. Charleston, SC: Arcadia Publishing, 2001.

———. *Deep River and Ivoryton*. Charleston, SC: Arcadia Publishing, 2002.

Manstan, Roy, and Frederic Frese. *Turtle: David Bushnell's Revolutionary Vessel*. Yardley, PA: Westholme Publishing, 2010.

McKay, Richard C. *South Street, A Maritime History of New York*. New York: G.P. Putnam's Sons, 1934.

Middlebrook, Louis F. *History of Maritime Connecticut During the American Revolution*. Vol. 1. Salem, MA: Essex Institute, 1925.

Payne, Ralph D. *The Old Merchant Marine*. New Haven, CT: Yale University Press, 1919.

Roberts, Jerry. *The British Raid on Essex: The Forgotten War of 1812*. Middletown, CT: Wesleyan University Press, 2014.

Stevens, Thomas A. *Connecticut River Master Mariners*. Essex, CT: Connecticut River Foundation at Steamboat Dock, 1979.

Story, Dana A. *The Shipbuilders of Essex (MA)*. Gloucester, MA: Ten Pound Island Book Company, 1995.

Tales of the Dauntless Club. Essex, CT: privately published, 1964.

Urquhart, William W. *Reminiscences: The Merchant Marine*. New York: Knickerbocker Press, 1910.

Articles

Harfst, Verena. "A Brief History of Native Americans in Essex, Connecticut." Essex Historical Society, 2019.

Milkofsky, Brenda. "Full Steam Ahead: Steamboat Travel in Connecticut." *Connecticut Explored* 7, no. 2 (Spring 2009).

Pagliuco, Christopher. "Ivoryton." *Connecticut Explored* 6, no. 1 (Fall 2008).

Family Records/Primary Sources

Five generations of Post/Rockwell/Urquhart/Collins family Bibles

Mayflower Lineage Records—John Alden/Priscilla Mullen through Ingham Line

Mayflower Lineage Records—John Howland and Elizabeth Tilley through Denison Line

Personal papers of Don Malcarne regarding New City

Post Family Genealogy by Robert Denison

Post Family Register of Captain David and Lucretia Post

Post/Hale Fan Lineage Chart

Post, Lewis. *Ship Journal of a Voyage to Hong Kong on the Ship Good Hope, 1869*.

Post/Rockwell Fan Lineage Chart

Whitney Family Register

Journals, Pamphlets, Magazines, Personal Communications

Connecticut River Museum. *Earl Brockway: Connecticut Boat Builder*.

Davis, Caryn B. "Seth Perrson." Personal communication, 2019.

Essex Historical Society. "Essex Connecticut: Where History Lives." 2015.

Essex Land Trust and Essex Historical Society. "Follow the Falls." 2018.

Hall, Jim. "Art Finkeldey." Personal communication, 2019.

Jones, Stephen. "CT River Boatbuilders." Personal communications, 2019.

Libby, Ned, and Strick Hyde. "*Chaco*." Personal communication, 2019.

Milkofsky, Brenda. "The Rise of Power Boating on the Connecticut River." *Connecticut Explored*, Summer 2018.

National Fisherman Yearbook. "They Don't Make Boatbuilders Like Earl Brockway Anymore." 1984.

Santana. "Schooner Dauntless." September 2005.

A Statement of the Facts and Circumstances Relative to the Operations of the Pilot Laws of U.S. With Particular Reference to New York. New York: R.C. Root and Anthony, 1848.

Maps/Surveys

Connecticut River Museum and Essex Historical Society. "A Walking Map of Essex CT."

F.W. Beers and Company. "A Map of Essex Village." New York, 1874.

H.&C.T. Smith and Company. "Essex Business Directory Map." New York, 1859.

LFW Land Surveying. "General Location Surveys, Title Sketches, New City Highway to the Landing at New City Wharf, Circa 1803, 1819, 1855, 1875." New City, Essex, Connecticut, 2005.

Websites

Battle Site Essex. www.battlesiteessex.org

Connecticut Explored. www.ctexplored.org.

Connecticut River Museum. www.ctrivermuseum.org.

Deep River Historical Society. "Town of Deep River, Connecticut Historic and Architectural Resources Survey. Phase II, 2012–2013." www.deepriverhistoricalsociety.org.

Essex Historical Society. www.essexhistory.org.

Essex Land Trust. www.essexlandtrust.org.

The Griswold Inn. www.griswoldinn.com.

Mount Saint John School. mountsaintjohnsschoolalumni.blogspot.com.

National Building Arts Center. www.nationalbuildingarts.org.

National Museums Liverpool. www.liverpoolmuseums.org.uk.

NavSource Naval History. www.navsource.org.

New England Lighthouses: A Virtual Guide. www.newenglandlighthouses.net.
Newspapers.com.
The Sound School. http://sound.school.
Town of Essex. www.essexct.gov.

Index

W

About the Authors

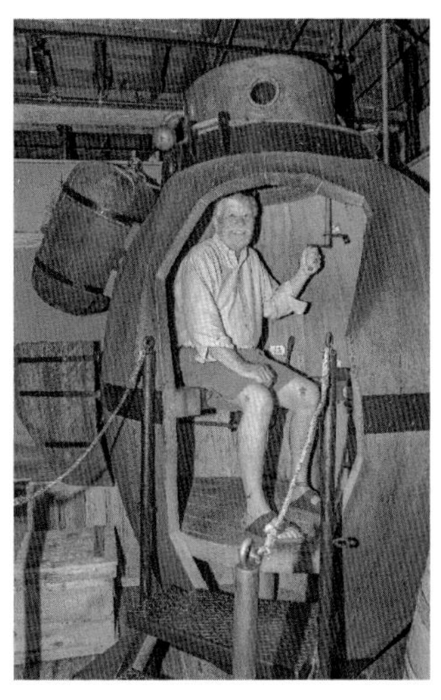

WICK GRISWOLD is the author of several History Press books. He teaches Sociology of the Connecticut River at the University of Hartford. He is also the commodore of the Connecticut River Drifting Society.

A former educational program director and regional high school teacher, RUTH MAJOR grew up hearing stories about her Saybrook/Essex ancestors who were shipmasters and shipbuilders. She credits her grandmother Marjorie Post for inspiring her passion for New England and New York history. Ruth lives on the Vineyard, where she researches, writes and paints.